Defects and Damage in Composite Materials and Structures

Rikard Benton Heslehurst

CRC Press
Taylor & Francis Group
Boca Raton   London   New York

CRC Press is an imprint of the
Taylor & Francis Group, an **informa** business

CRC Press
Taylor & Francis Group
6000 Broken Sound Parkway NW, Suite 300
Boca Raton, FL 33487-2742

First issued in paperback 2017

Version Date: 20140114

ISBN 13: 978-1-4665-8047-3 (hbk)
ISBN 13: 978-1-138-07369-2 (pbk)

Library of Congress Cataloging-in-Publication Data

Heslehurst, Rikard Benton.
    Defects and damage in composite materials and structures / author, Rikard Benton Heslehurst.
        pages cm
    Includes bibliographical references and index.
    ISBN 978-1-4665-8047-3 (hardback)
    1. Composite materials--Defects. I. Title.

TA418.9.C6H48 2014
620.1'18--dc23   2013049509

Visit the Taylor & Francis Web site at
http://www.taylorandfrancis.com

and the CRC Press Web site at
http://www.crcpress.com

I dedicate this book to my family (my wife Fiona and our three wonderful children) and in particular to my parents (Rodde and Janice Heslehurst).

Contents

List of Figures

List of Tables

Nomenclature

Acronyms

ACSS	Australian Composite Structures Society
AFRP	aramid fiber reinforced plastic
BFRP	boron fiber reinforced plastic
BVID	barely visible impact damage
CFRP	carbon (graphite) fiber reinforced plastic
EAA	Experimental Aircraft Association
FB	fiber breakage
FS/DEL	fiber shear/delamination
GFRP	glass fiber reinforced plastic
NDI	nondestructive inspection
SAE	Society for Automotive Engineers

Symbology

D_{hole}	= hole diameter
$E_{damage1}$	= damaged region stiffness
$E_{parent1}$	= parent laminate effective principle stiffness
E_{patch1}	= patch stiffness
h	= laminate thickness
L_{patch}	= patch length
t_{damage}	= depth of damaged region
t_{patch}	= patch thickness
t_{repair}	= thickness of repaired region

Preface

This book aims to provide a detailed description of defects and damage associated with advanced composite materials and structures. The topic of defects and damage in advanced composite materials has been discussed in many reports and articles over the past 50 years. Most of the early work during the 1970s through to the 1990s has not lost any relevance in today's technological understanding of composite materials and structures. In fact, current work in the field of defects and damage only confirms the understanding and outcomes of the earlier work to be relevant and applicable. This book has accumulated the information developed over the past decades and organized this data and knowledge into a concise document. The author's own experience in the field of in-service activities for composite airframe maintenance and repair is added to that wealth of knowledge, with a particular thrust into the criticality of damage from a structural integrity point of view, the relevance of identifying the damage, and how to repair the defects and damage.

Chapter 1 outlines the general applications of composite materials and fundamental definitions of defects and damage. Chapter 2 provides a description of the defects and damage types in composite materials and structures. This chapter categorizes the defects and damage types based on size, occurrence, and location, leading to generalization of the types of defects and damage. Chapter 3 provides guidance into the various NDI methods that can be used to find and identify the defects and damage at various levels of detail and description. Chapter 4 reviews the basic failure modes and mechanisms of composite materials defects. This chapter leads to the determination of the stress state effects on the composite structure of such defects in Chapter 5. In Chapter 5 the loss of structural integrity is determined from the generalized defect types of matrix cracks, delaminations, and fiber fractures. Recommendations are provided for determining the stress state and defect/damage criticality to the functional performance of the composite materials and structure. Finally, Chapter 6 describes the recommended restoration practices for defect and damage repair action.

Acknowledgments

The work described in this book would not have been possible except with the assistance from the following people and organizations:

- Mike Hoke and the team at Abaris Training
- Staff at University College (UNSW), at the Australian Defence Force Academy
- The several supervisors I had during my time with the Royal Australia Air Force
- Professor Rhys Jones during his time at the Australian Defence Science and Technology Organisation

About the Author

Rikard Benton Heslehurst, PhD,
is a former aeronautical engineering officer in the Royal Australian
Air Force (RAAF). During his 16
years of military service, Rik was
in charge of the RAAF Material
and Process Engineering Section
and earlier an airworthiness engineer on the F/A-18 Hornet aircraft.
Currently, Rik is a senior lecturer
in the School of Engineering and
Information Technology of the University College, University
of New South Wales (UNSW), the Australian Defence Force
Academy (UNSW@ADFA). At UNSW@ADFA, Rik lectures in
aircraft design, airframe design and analysis, structural joining methods, damage analysis and repair design, and composite structural design. His research interests follow similar
lines. Rik is currently the senior engineer for Abaris Training
in Reno, Nevada, and he consults for the Australian Defence
Force, Civil Aviation Safety Authority, Raytheon Missile
Systems, NASA, USAF, Boeing Airplane Company, Bombardier
Aerospace, Australian Space Safety Office, SRAM-Zipp Wheels,
and SP Systems-UK. He has conducted engineering short
courses for NASA's Kennedy, Marshall, and Goddard Space
Flight Centers, Lockheed-Martin Skunk Works, USAF Academy,
Royal Australian Navy, USAF Wright Patterson Laboratories,

Pratt & Whitney CT, USAF Warner Robins Air Logistics Center, Colombian Air Force, Boeing Australia, Raytheon-Australia, Bombardier Aerospace, Honeywell Systems, Sandia National Laboratories, General Dynamics, Pratt & Whitney, Gulfstream Aerospace, SRAM-Zipp Wheels, and the Republic of Singapore Air Force.

Rik earned a bachelor of engineering (aeronautical) degree, with honors, and a master of engineering degree, both from the Royal Melbourne Institute of Technology, and has also earned a PhD from the UNSW. Rik is the current chair of the Australian Chapter of the Society for the Advancement of Materials and Process Engineering (SAMPE) and chair of the Canberra Branch of the Australian Composite Structures Society. He has also been involved with the Canberra Branch of the Royal Aeronautical Society and the Canberra Division of Engineers Australia.

Rik is a chartered professional engineer; a fellow of the Institute of Engineers, Australia; a fellow of the Royal Aeronautical Society; an elected SAMPE fellow; and a senior member of the American Institute of Aeronautics and Astronautics. He is also a member of SAE, ACSS, EAA, and the Spitfire Association and an honorary member of Composites Australia.

Chapter 1

Introduction

The application of advanced materials in components and structures has evolved due to the need to reduce structural weight and improve performance. Other attributes of composite materials, such as corrosion resistance, excellent surface profiles, enhanced fatigue resilience, and tailored performance, have also been significant contributors to the rapid rise in composite materials application. As a result, these new materials are required to perform at higher stress levels than previous applications while also providing adequate levels of damage tolerance. Advanced composite materials provide the necessary damage tolerance through relatively low, applied design strains. However, defects and damage still occur in composite materials, and it is the assessment of defect and damage criticality and the subsequent repair requirements that are currently challenging for operators of composite materials.

When composite materials components are damaged or defective in some way, the engineer/technician needs to determine the size, shape, depth, type, and extent of the anomaly and restitution approach. A typical repair procedure is shown in Table 1.1. Of immediate importance is the ability to identify the damage and determine its extent by some suitable nondestructive inspection (NDI) technique. Most, if not all,

1

Table 1.1   Typical Composite Structure Repair Procedure

1.	Locate the damaged area
2.	Assess the extent of damage
3.	Evaluate the stress state of the damaged area stress state
4.	Design the repair scheme
5.	Remove damage and repair structure
6.	Fabricate and prepare the repair scheme
7.	Apply the repair scheme
8.	Conduct post-repair quality checks
9.	Document repair procedures
10.	Monitor the repair region

of the standard NDI techniques currently used require high levels of operator experience to successfully apply the NDI technique and interpret the results.

This book is written to provide an in-depth study of defects and damage in composite materials. It is significantly focused on the defect and associated structural response to the presence of defects.

What Does This Book Contain?

Chapter 2 describes damage and defect types in detail. Both a written and (where possible) an illustrative description of the many types of defects and damage in composite materials are given. The discussion leads to a consideration of how defects and damage can be categorized.

Chapter 3 provides a relatively brief overview of the various methods of finding defects and damage through nondestructive methods of inspection (NDI). Finding the nonconformity can be a challenge, and Chapter 3 provides a tabulate guide of the defect-discovery capability of the various NDI methods discussed.

Chapter 4 lists the failure mechanisms of the generalized defects and damage types. The chapter describes important relationships to load type and orientation of the defect and damage.

Chapter 5 covers the loss of structural or performance integrity based on the outcome of Chapter 4. This chapter will allow a better appreciation of the repair requirements needed to restore the composite structure/material to the level needed.

Chapter 6 provides a short overview of the principal repair methods and processes for restitution of defective or damaged composite components.

Definitions

The following define what constitutes a defect or damage and what constitutes failure of materials, components, and structures (Heslehurst 1991):

Defects: A material or structural defect, which is also known as a discontinuity, flaw, or damage, is defined as "any unintentional local variation in the physical state or mechanical properties of a material or structure that *may* affect the structural behavior of the component." The word *may* is used in the definition, as local nonconformities do not necessarily adversely affect material or structural performance.

Failure: The failure of a component or structure is defined as "when a component or structure is unable to perform its primary function adequately." The primary function may not be just structural performance (strength or rigidity), but also environmental resistance, electrical and thermal conductivity, energy absorption, etc.

Advanced Composite Materials

The application of advanced composite materials in primary load-bearing structures is steadily increasing. The term *advanced* refers to those new fiber/resin composite systems that have greater strength and stiffness properties over conventional glass fiber material. Composite materials provide many significant advantages over conventional metals used in structures applications; importantly, advanced composite materials have excellent specific strength and specific stiffness properties. Advanced composite materials also possess some distinct attributes, such as the fact that the fibers and resin remain individual constituents, with the fibers providing the strength and stiffness and the resin protecting the brittle fibers and providing load transfer between adjacent fibers. However, it is the design freedom that gives composite materials their greatest advantages and uniqueness.

Advanced two-dimensional (laminated) composite materials have their limitations in structural applications. The materials are planar in nature. That is, the fibers are all in one plane, and the technique of lamination means that the principal structural properties are also in the same plane (see Figure 1.1). As a result, the through-the-thickness mechanical properties are significantly weaker than the in-plane properties. This has meant that composite structures are essentially designed for only in-plane loads. However, most damage in composite structures is associated with impact and the introduction of out-of-plane stresses. It is these out-of-plane stresses (known as interlaminar stresses) that ultimately cause a loss of composite component structural integrity. Understanding such

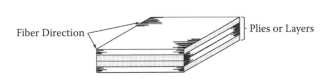

Figure 1.1   Composite laminate showing its planar nature.

damage is an area of major research. This research attempts to better understand damage initiation and propagation in advanced composite materials (Wang, Slominana, and Bucinell 1985; Williams et al. 1986; Lessard and Liu 1992; Lagace and Bhat 1993; Chen and Chang 1994).

The application of advanced composite materials in airframe structures is increasing with every new aircraft designed and produced. This is particularly so with military fighter aircraft. The F-15 Eagle, a 1968 aircraft, has 1.5% of its structural weight made from advanced composite materials. The F-18 Hornet, which commenced manufacture in 1974, has 12.1% of its structural weight made from advanced composite materials. Advanced composite materials make up 26% of the structural weight of the AV-8B Harrier II (1978), and the F-22 has about 24%, with the F-35 comprising 38%. A typical breakdown of the structural materials used in a modern airframe is shown in Figure 1.2. Rotor-wing airframes have exceeded the fixed-wing aircraft with composite materials. For example, the V-22 Osprey utilizes 65% of the airframe with composite materials; the Eurocopter Tiger

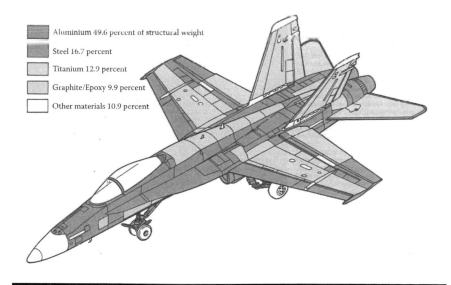

Aluminium 49.6 percent of structural weight

Steel 16.7 percent

Titanium 12.9 percent

Graphite/Epoxy 9.9 percent

Other materials 10.9 percent

Figure 1.2 (See color insert.)   Material breakdown of the F-18 Hornet airframe (United States Navy, 1984).

armed reconnaissance helicopter has 80%; and the Eurocopter NH-90 has 90% of carbon (graphite) fiber/resin composites. The commercial aircraft field has also had a significant increase of composite materials in the airframe from 24% in the A380, 52% in the A350, and 50% in the B787. Figure 1.3 provides a chart showing the increase in the use of composite materials in aircraft (military, helicopter, and civil) over the last few decades.

A composite material in the context of this book is a material consisting of any combination of filaments and/or particulates in a common matrix. Various material combinations can therefore be called composites, e.g., fiber-reinforced plastics (FRP), timber, and concrete (Figure 1.4).

The basic premise of the term *composite materials* is that the combination of different materials to form a new material is done such that each constituent material does not lose its individual form or material properties. The composite material is thus such a combination that each constituent material

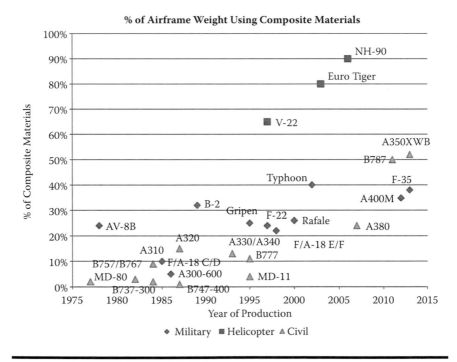

Figure 1.3   Growth in composite materials used in airframes.

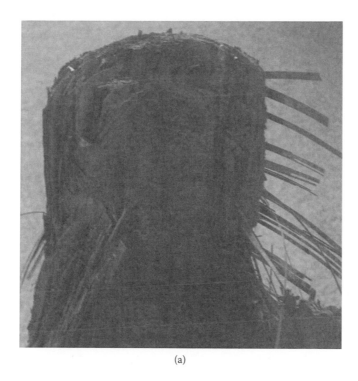

(a)

Figure 1.4a   Examples of three types of composite materials: FRP.

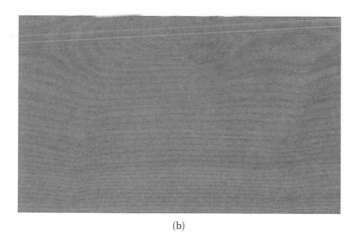

(b)

Figure 1.4b   Examples of three types of composite materials: timber.

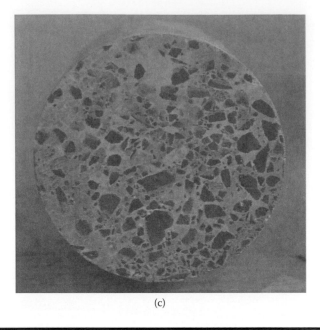

(c)

Figure 1.4c Examples of three types of composite materials: concrete.

maintains its own characteristics (both good and poor properties) but provides support to the other constituent to overcome limitations in the other constituent materials.

In this book, we are predominantly interested in the fiber-reinforced types of composite materials, such that the combined properties of the fiber and matrix are used to enhance one another. The filament or fiber or fabric (Figure 1.5a) provides the essential axial high strength and stiffness of the composite material, with a low density that gives the significant benefits of exceptionally high specific properties. However, the filament, fiber, or fabric is brittle and requires support against premature fracture. Thus a matrix is added to the fiber (polymer resin, ceramic, or metal) (Figure 1.5b) to provide good shear behavior with an ease of fabrication and with a relatively low density. The material is generally more susceptible to defects and damage from the operational environment. Thus a composite material (i.e., structure) is formed by the

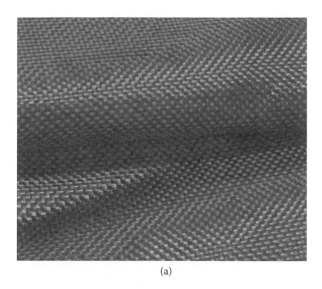

(a)

Figure 1.5a   Constituents and composite material: fibers.

combination of the fiber and matrix constituents (Figure 1.5c). The resulting structure provides an increase in the damage tolerance and toughness of the brittle fibers with minimal loss of their beneficial mechanical and physical properties.

The general classification of composite materials is listed in Table 1.2.

With respect to current composite industry applications, we will confine our discussions to ceramic/polymer composites such as graphite (carbon)/resin (CFRP) and glass/resin (GFRP), and polymer/polymer composites such as aramid/resin (AFRP). The use of boron/resin (BFRP) composites is typically reserved for a special range of higher strength and stiffness applications, particularly for compression.

Range of Composite Structural Types

The application of composite materials for structural purposes has been around for a century or more. Apart from the use of natural composites (timber), composite materials

(b)

Figure 1.5b   Constituents and composite material: matrix.

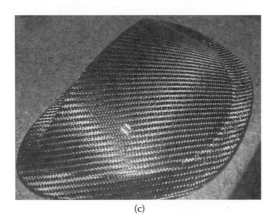

(c)

Figure 1.5c   Constituents and composite material: structure.

Table 1.2 Classification of Composite Materials

Matrix	Reinforcement: Fiber (F) or Particulate (P)		
	Metal	*Ceramic*	*Polymer*
Metallic	*F/P*	*F/P*	N/A
Ceramic	*F*	*F(P?)*	*F*
Polymer	*F*	***F/P***	***F***

Note: Bold letters indicate the focus of this book.

Figure 1.6 (See color insert.)   Wright Military Flyer B Model (1909).

Figure 1.7 (See color insert.)   Supermarine Spitfire (1940).

in their current advanced form (fiber-reinforced polymer resins) have only been in use for about six decades. This is primarily due to the development of the advanced fibers. This historical application of composite materials is illustrated in Figures 1.6 to 1.9.

Figure 1.8 (See color insert.)   de Havilland DH98 Mosquito (1938).

Figure 1.9 (See color insert.)   Lockheed P-2V Neptune (1947).

Aerospace Applications

For airframes, the design requirements for joining composite structures can be summarized as:

- High strength-to-weight ratio
- High stiffness-to-weight ratio
- Damage tolerance
- Corrosion resistance
- Manufacturing efficiency
- Structural and operational effectiveness and efficiency
- Structural and system compatibility
- Maintainability and repairability

Current airframe designs are steadily increasing the volume of composite materials in use and the joining processes are becoming of greater importance. Several applications of composite joints are illustrated in the following aircraft:

Boeing/MDD Harrier (Figure 1.10)
- Wing skins and substructure
- Empennage skins and substructure
- Fuselage skins
- Forward fuselage substructure

Boeing/Bell V-22 Osprey (Figure 1.11)
- Entire wing, empennage, and fuselage structure
- Rotor blades

Airbus A380 (Figure 1.12)
- Empennage skins and some substructure
- Wing skins
- Flight control structure
- Access panels and doors
- Engine nacelle

Figure 1.10 (See color insert.)   Boeing/MDD Harrier.

Figure 1.11 (See color insert.) Boeing/Bell V-22 Osprey.

Figure 1.12 (See color insert.) Airbus A380.

Boeing/MDD C-17 Globemaster III (Figure 1.13)

■ Fuselage structure
■ Empennage structure
■ Wing skins
■ Access panels
■ Undercarriage doors and wheel pod
■ Engine nacelle

Figure 1.13 (See color insert.)   Boeing/MDD C-17 Globemaster III.

Figure 1.14 (See color insert.)   Sports aircraft.

Sports aircraft (Figure 1.14)
- Entire surface structure
- Wing spar
- Propeller
- Wheel pants

X-34 (NASA) (Figure 1.15)
- Entire structure
- Fuel tanks

Figure 1.15 (See color insert.)   X-34 (NASA).

Figure 1.16 (See color insert.)   Wind turbines.

Civil Infrastructure Applications

The basic design requirements for the composite civil infrastructure can be summarized as follows:

Wind turbines (Figure 1.16)
- ■ Environmental and corrosion resistance
- ■ Durability and longevity
- ■ Damage tolerance
- ■ Simple and effective manufacturing methods

■ Efficient designs
■ Vibration and seismic considerations
■ Improved performance and cost benefits

Marine Applications

For marine applications, the design requirements are as follows:

Luxury cruiser (Figure 1.17)
■ Saltwater environment
■ Slamming loads
■ Non-autoclave curing
■ Long working life

Land Transport Applications

Composite design requirements:

Auto racing (Figure 1.18)
■ Cost efficiency
■ Manufacturing simplicity
■ Performance (weight, strength)
■ Durability and damage tolerance

Figure 1.17 (See color insert.) Luxury cruiser.

Figure 1.18 (See color insert.)   Auto racing.

Figure 1.19   High-performance sporting equipment.

Other Applications

Design requirements for composite components:

High-performance sporting equipment (Figure 1.19)
- ■ Economical design
- ■ Comfortable and lightweight structure

- Durability
- Environmental resistance
- Impact resistance
- High performance

Design Requirements with a Focus on the Damaging Environment

The basic design requirements for the development of composite components and structure can be considered as follows:

- Performance:
  - Functional performance:
    - Tensile, compressive, shear, and/or bearing strength of the composite components and structures
    - Strength loss of composite components and structures due to defects and damage
    - Out-of-plane strength of composite components and structures with and without defects
    - Young's Modulus in the orthogonal directions of composite components and structures
    - Axial Young's Modulus of fasteners
    - Stiffness loss of composite components and structures due to defects and damage
    - Coefficient of thermal expansion of composite components and structures with and without defects
    - Coefficient of moisture absorption of composite components and structures with and without defects
  - Spatial constraints:
    - Defect or damage size (three dimensional)
    - Relative defect or damage size to component/structure size
    - Density of defect or damage
    - Location of defect or damage on or in the structure

- Appearance:
  - Effect of defect or damage to the surface profile of the component or structure
- Time:
  - Time to identify defect or damage
  - Time to repair defective or damage structure
  - Time over which defect or damage has manifested itself
- Cost:
  - Cost of identification of defect or damage
  - Cost of repairing the structure
  - Component or structure downtime cost impact
- Manufacture/assembly:
  - Repair capability
- Standards:
  - Defect representation
  - NDI personal and equipment
  - Structural repair manual guidance
- Safety of personnel:
  - During damage inspection
  - During damage repair
- Environmental issues:
  - Environmental impact due to the existence of the damage or defect
- Maintenance and repair:
  - Component/structure access
  - Component/structure inspectability
  - Component/structure repairability
  - Facility capabilities
- Personal training

The application of Quality Function Deployment methodology for the development of function design specification (Ullman 2009) will assist in understanding the impact of defects and damage on composite components and structures.

Chapter 2

Damage and Defect Description

Introduction

Defects and damage in structural components are common occurrences, whether they arise during material processing, component fabrication, or in-service action. The effect of the defect or damage in the composite component's structural integrity is essential in understanding the criticality of the defect. This chapter first reviews all of the known defect and damage types in composites structure, with discussion and tabulation of when they occur in the composite structure, their general size of occurrence, and the typical location of said defects. From this study and discussion, the defects are then generalized into four categories.

Defect Types

There are some 52 separate defect types that composite components are prone to or potentially subjected to. They range from microscopic fiber faults to large, gross impact damage. In this chapter, each defect is described in some detail, and then these defects are categorized into specific groups that will assist in the identification of those defects that are of particular concern to the in-service life of composite and bonded aircraft structures.

Description

A detailed description of the 52 defect types that exist in composite structures is provided in Appendix A to this chapter. These 52 defects are listed in alphabetical order in Table 2.1.

Classifications of Defect Types

Defects can be grouped into specific categories according to when they arise during the life of the composite structure, their relative size, their location or origin within the composite structure, and their production of a similar effect to a known stress state in the composite component.

Defect Occurrence

Defects occur during materials processing, component manufacture, or in-service use. Table 2.2 lists those defects peculiar to manufacturing (materials processing and component manufacture) and those sustained during service life.

Materials processing. Materials-processing defects occur during the production and preparation of the constituent materials of a prepreg (preimpregnated composite fibers) because of improper storage or quality control and batch certification procedures leading to material variations.

Component manufacture. Component manufacture-induced defects occur during either the layup and cure or the machining and assembly of the components.

In-service use. In-service components will have defects that occur through mechanical action or contact with hostile

Table 2.1   Composite Material Defect Types

Bearing surface damage	Blistering	Contamination
Corner crack	Corner/edge splitting	Corner radius delaminations
Cracks	Creep	Crushing
Cuts and scratches	Damaged filaments	Delaminations
Dents	Edge damage	Erosion
Excessive ply overlap	Fastener holes	Fiber distribution variance
Fiber faults	Fiber kinks	Fiber/matrix debonds
Fiber misalignment	Fracture	Holes and penetration
Impact damage	Marcelled fibers	Matrix cracking
Matrix crazing	Miscollination	Mismatched parts
Missing plies	Moisture pickup	Nonuniform agglomeration of hardener agents
Over-aged prepreg	Over/under cured	Pills or fuzz balls
Ply underlap or gap	Porosity	Prepreg variability
Reworked areas	Surface damage	Surface oxidation
Surface swelling	Thermal stresses	Translaminar cracks
Unbond or debond	Variation in density	Variation in fiber-volume ratio
Variation in thickness	Voids	Warping
Wrong materials		

Table 2.2   Listing of Manufacturing and In-Service Defects

Materials Processing	Component Manufacture	In-Service
Damaged filaments	Blistering	Bearing surface damage
Fiber distribution variance	Contamination	Corner/edge crack
	Corner/edge splitting	Corner radius delamination
	Cracks	Creep
Fiber faults	Delaminations	Crushing
Fiber/matrix debonds	Debond	Cuts and scratches
	Excessive ply overlap	Delaminations
Fiber misalignment	Fastener holes	Debond
	• Elongation	Dents
Marcelled fibers	• Improper installation	Edge damage
Miscollination	• Improper seating	Erosion
Over-aged prepreg	• Interference fitted	Fastener holes
	• Missing fasteners	Elongation
Prepreg variability	• Overtorqued	Hole wear
	• Pull-through	Improper installation
	• Resin-starved bearing surface	Improper seating
		Interference fitted
	• Tilted countersink	Missing fasteners
	Fiber kinks	Overtorqued
	Fiber misalignment	Pull-through
	Fracture	Removal and reinstallation
	Holes	
	• Drill burn	Tilted countersink
	• Elongation	Fiber kinks
	• Exit delamination	Fracture
	• Misdrilled and filled	Holes and penetration
	• Porosity	Matrix cracking
	• Tilted	Matrix crazing

Table 2.2 (*Continued*)   Listing of Manufacturing and In-Service Defects

Materials Processing	Component Manufacture	In-Service
	Mismatched parts	Moisture pickup
	Missing plies	Reworked areas
	Nonuniform agglomeration of hardener agents	Surface damage
		Surface oxidation
	Over/under cured	Surface swelling
	Pills and fuzz balls	Translaminar cracks
	Ply overlap or gap	
	Porosity	
	Surface damage	
	Thermal stresses	
	Variation in density	
	Variation in resin fraction	
	Variation in thickness	
	Voids	
	Warping	
	Wrong materials	

environments, such as impact and handling damage, local overloading, local heating, chemical attack, ultraviolet radiation, battle damage, lightning strikes, acoustic vibration, fatigue, or inappropriate repair action.

Defect Size

The size of a defect has significant bearing on its criticality. Therefore, composite defects are listed under the levels of size—microscopic and macroscopic—in Table 2.3.

Table 2.3 Listing of Defects by Relative Size

Microscopic	*Macroscopic*
Contamination	Bearing surface damage
Creep	Blistering
Damaged filaments	Contamination
Fiber distribution variance	Corner cracks
Fiber faults	Corner/edge splitting
Fiber/matrix debonds	Corner radius delamination
Fiber misalignment	Cracks
Marcelled fibers	Crushing
Matrix cracking	Cuts and scratches
Miscollination	Debond
Moisture pickup	Delaminations
Nonuniform agglomeration of hardener agents	Dents
	Edge damage
Over/under cured	Erosion
Pills and fuzz balls	Excessive ply overlap
Prepreg variability	Fastener holes
Surface oxidation	Fracture
Thermal stresses	Holes and penetration
Variation in density	Matrix cracking and crazing
Variation in resin fraction	Mismatched parts
Variation in thickness	Missing plies
Voids	Ply overlap or gap
Wrong materials	Porosity
	Reworked areas
	Surface damage
	Surface swelling
	Translaminar cracks
	Variation in thickness

Table 2.3 (*Continued*)   Listing of Defects by Relative Size

Microscopic	*Macroscopic*
	Voids
	Warping
	Wrong material

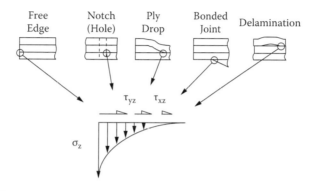

Figure 2.1   Sources of out-of-plane loads.

Defect Location

Defects may be present in isolation originating from structural features such as cutouts and bolted joints, or a random accumulation resulting from interaction between other defects. However, they tend to concentrate at geometric discontinuities, as illustrated in Figure 2.1. Under the headings of geometric discontinuities, free edges, projectile impact, and heat damage, the defects are classified in Table 2.4 as location of in-service defects.

Generalization of Defect Types

The results of a literature survey (Heslehurst and Scott 1990) indicate that defects can be listed in terms of developing a common stress state. These common stress states are delaminations, transverse matrix cracks, holes or fiber fracture, and design variance (see Table 2.5).

Table 2.4 Location of In-Service Defects

Geometric Discontinuities	Free Edges	Projectile Impact	Heat Damage
Corner/edge crack	Bearing surface damage	Crushing	Creep
Corner radius delamination	Delaminations	Cuts and scratches	Matrix cracking
Debond	Edge damage	Debond	Matrix crazing
Fastener holes	Erosion	Delaminations	Surface damage
Fiber kinks	Fastener holes	Dents	Surface oxidation
Reworked areas	Holes and penetration	Fracture	Surface swelling
	Moisture pickup	Holes and penetration	
		Surface damage	
		Translaminar cracks	

Conclusions

Advanced polymeric composite materials are prone to a large number of defects and damage types. These defects and damage types emanate from the constituent material processes, composite component manufacture, and in-service use of the composite component. Those defect types due to in-service usage can be generalized further as:

a. Transverse matrix cracks
b. Delaminations
c. Holes (fiber fracture)

Table 2.5   Generalized Defect Types

Delaminations	Matrix Cracks	Holes	Design Variance
Bearing surface damage	Bearing surface damage	Bearing surface damage	Creep
Blistering	Contamination	Crushing	Damaged filaments
Contamination	Corner/edge crack	Cuts and scratches	Dents
Corner/edge crack	Cracks	Fastener holes	Erosion
Corner radius delamination	Edge damage	Fiber kinks	Excessive ply overlap
Debond	Matrix cracking	Fracture	Fiber distribution variance
Delaminations	Matrix crazing	Holes and penetration	Fiber faults
Edge damage	Porosity	Reworked areas	Fiber kinks
Fastener holes	Translaminar cracks	Surface damage	Fiber misalignment
Fiber/matrix debond	Voids		Marcelled fibers
Holes and penetration			Miscollination
Pills and fuzz balls			Mismatched parts
Surface swelling			Missing plies
			Moisture pickup
			Nonuniform agglomeration of hardener agents
			Over-aged prepreg
			Over/under cured
			Pills and fuzz balls

(Continued)

Table 2.5 (*Continued*) Generalized Defect Types

Delaminations	Matrix Cracks	Holes	Design Variance
			Ply underlap/ gap
			Prepreg variability
			Surface oxidation
			Thermal stresses
			Variation in density
			Variation in resin fraction
			Variation in thickness
			Warping
			Wrong materials

The failure modes of composite materials are numerous and are influenced by many factors. In multidirectional laminates, the prediction of failure modes is very difficult and is usually a combination of several unidirectional failure modes under the various loading spectrums. In the majority of failure cases, the failure mode is determined by postmortem examination of the fracture surface.

Appendix A: Defect Type Description

The following definitions provide a detailed description of all of the defect types that are likely to occur in a composite component.

All these defects will affect the bearing strength to varying degrees.

Bearing surface damage: Occurs at the contact point between a pin (fastener) and the hole edge. The damage is likely to contain fiber fracture, delaminations, and matrix cracking, and is a result of improper fastener installation, joint overload, or loose fasteners (Figure A2.1). Bearing surface damage is known to reduce the joint stiffness and bearing/bypass load response. Bearing surface damage can be treated as delaminations and cracks.

Blistering: Localized lamina (ply) delaminations (Figure A2.2). Blistering can occur anywhere in the lamina and is caused by the expansion of trapped gases within the lamina. Surface blisters can occur

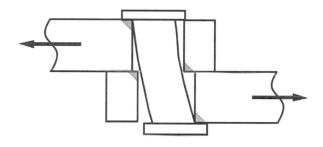

Figure A2.1   Primary source of bearing failure.

Figure A2.2   Localized surface blister.

due to chemical attack or localized heating of the matrix. For the effect blistering has on a laminate, see **delaminations**.

Contamination: The inclusion of foreign materials in the laminate such as peel ply or backing paper, usually between plies, during fabrication of the component. Depending on the size and extent of the contamination, there will be a varying effect on the component. Contaminations can be represented as **delaminations**.

Corner crack: Matrix crack, either perpendicular or trans-laminar to the ply (Figure A2.3). Cracks are caused by the same reasons as **corner splitting**. See **cracks** and **delaminations**.

Corner/edge splitting: An edge delamination typically due to edge impact (Figure A2.4). These delaminations are cracks between plies that run parallel to the ply interface as depicted in Figure A2.5. Edge delaminations are caused by out-of-plane stresses generated at the edge or through impact damage, usually

Figure A2.3 Corner radius cracks.

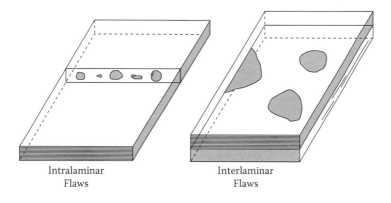

Intralaminar
Flaws

Interlaminar
Flaws

Figure A2.4   Corner edge damage from localized impact.

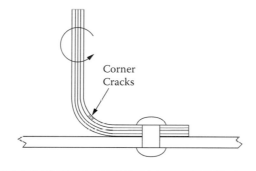

Corner
Cracks

Flgurc A2.5   Effectlve intralaminar and interlaminar flaws.

maintenance related. Splitting is detrimental to shear
strength. See **delaminations**.

Corner radius delaminations: Matrix cracks running parallel to the fiber axis in the corner radius of a component, usually a stiffener (Figure A2.6). Predominantly a
manufacturing error, corner radius delaminations result
in out-of-plane stresses being induced into the component. The delamination will run longitudinally. See
delaminations.

Cracks: In terms of this work, cracks in a laminate are
those that occur within the matrix only; when a
fiber is cracked, it is referred to as fiber **fracture**.
Matrix cracks are characterized by localized partial

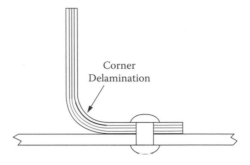

Figure A2.6   Corner radius delamination.

through-the-thickness cracking. Many cracklike failures in continuous fiber composites and their laminates are contained within the interface planes, where the matrix material properties dominate the fracture response. Cracks produce localized stress concentrations and, if severe, can lead to fiber fracture or **delaminations**. **Matrix cracking** is generated by overstressing of the matrix through various loading conditions and can occur at relatively low loads, even thermal expansion during the cure cycle. The various types of matrix cracks are illustrated in Figures A2.5 and A2.7. A micrograph of matrix cracking is shown in Figure A2.8.

Creep: The plastic deformations caused by sustained loading, usually at high temperatures. Creep is a matrix-dominated failure, and therefore it tends to affect compression and shear performance to a greater extent than tension. However, tests on typical structural elements have shown creep not to be a problem, and it is considered that there is no significant interaction between defects and creep. However, creep must be considered in size/shape critical components where in-plane extension is constrained and the resulting out-of-plane deflections may lead to buckling.

Crushing: Local indentations or surface dents caused by impact damage. It may be a sign that there is further

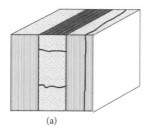

(a)

Figure A2.7a Intralaminar cracks.

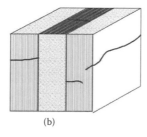

(b)

Figure A2.7b Translaminar fracture.

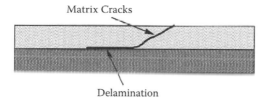

Figure A2.8 Matrix crack and delamination initiation.

internal damage, such as **delaminations**, fiber **fracture**, or **matrix cracking**. On the external surface of the component, matrix cracks and fiber fracture can be present with crushing. Crushing is more common in laminate/honeycomb core sandwich constructions, as shown in Figure A2.9.

Cuts and scratches: Cuts (Figure A2.10) and scratches can be treated as surface damage. The severity of surface scratches and notches depends on their width, depth, and orientation to the fibers or loading direction.

Figure A2.9   Crushing of composite sandwich panel.

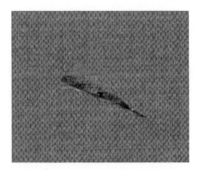

Figure A2.10   Composite surface cut.

A high reduction in the static strength is possible, but with the current design allowable strain they are not critical. See **cracks** and **fiber fracture**.

Damaged filaments: Broken filaments (Figure A2.11a), knots (Figure A2.11b), splices, split tow, fiber separation, hollow fibers, or interrupted fibers all come under the heading of damaged filaments. Such damage will reduce the filament strength and fiber/matrix interface strength, and thus could degrade the lamina stiffness. Filament damage is a result of poor prefabrication control and handling.

Delaminations: Also termed **interlaminar cracking**, delaminations are one of the most frequently encountered types of damage found in advanced composite

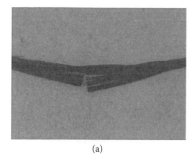

(a)

Figure A2.11a   Filament breakage in a tow.

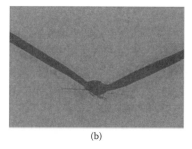

(b)

Figure A2.11b   Knot in a tow.

materials. Delaminations are a matrix defect, where in-plane matrix cracks propagate between plies of a laminate or within a laminate, where cracks run parallel to the fiber direction. Delaminations are graphically depicted in Figures A2.5, A2.8, A2.9, and A2.12. Delaminations may form and grow under static and cyclic tensile loading, but are predominantly a compression-related defect causing significant degradation to the component's compressive and shear strengths. Delaminations are caused by either:

a: Impact damage where internal interlaminar failure occurs

b: Free edges (geometric boundaries, microcracks, or voids), where the interlaminar stresses are high due to the mismatched Poisson's ratio

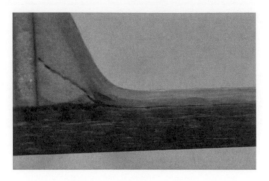

Figure A2.12   Delamination in composite structure.

   c: Arresting of through-the-thickness matrix cracks at
        a ply interface, after which the crack runs parallel
        to the interface, as shown in Figure A2.8

In the case of impact damage, all experimental evidence has shown that the visible surface damage is
considerably less than the actual internal damage
(ply delamination and fiber/matrix failure) or damage to the backside of the graphite/epoxy structure.
In all instances, ultrasonic inspection of the ballistic
entry points indicated larger areas of concealed damage than visible at the surface. Delaminations are
therefore often difficult to detect. The types of damage resulting from various impact energies are illustrated in Figure A2.13, which shows the effects on
a laminate due to high-, moderate-, and low-energy
impact projectiles. The laminate response to delaminations is influenced by the delamination size and
location, laminate orientation/stacking sequence, and
test environment. The larger the delamination and
the deeper it is located within the laminate, the larger
the strength loss. Only delaminations near the surface
grow in a stable manner, but these induce negligible
strength losses. Small delaminations only show negligible strength reductions. Delaminations induce local
interlaminar stresses.

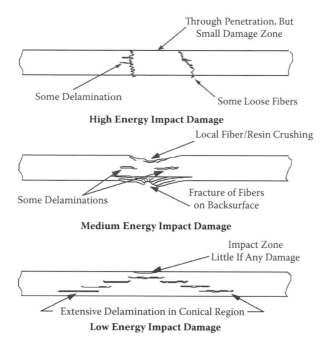

Figure A2.13 Impact damage at various energy levels.

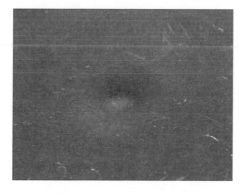

Figure A2.14 Surface dent from impact.

Dents: Indentations at the point of contact of a moving body. They are distinct from **crushing** because no fibers are broken (Figure A2.14). However, dents can result in loss of localized stiffness. Tool impressions are a common source of dents.

Edge damage: Caused by the mishandling of components. Common features of edge damage are splitting and delaminations. Edge delaminations can arise from high out-of-plane normal or shearing stresses produced in the vicinity of free edges, as shown in Figure A2.4 and illustrated in Figure A2.15. For a further description, see **delaminations**.

Erosion: The removal of surface material through wear and abrasion is called erosion (Figure A2.16). During erosion, outer matrix material and fiber are effectively removed. This will cause localized strength and stiffness degradation and produce an area of asymmetry, where out-of-plane stresses may be induced; this degradation effect will depend on the depth of erosion.

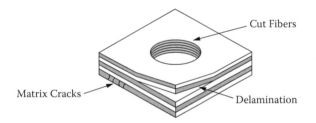

Figure A2.15 Principal in-service damage types.

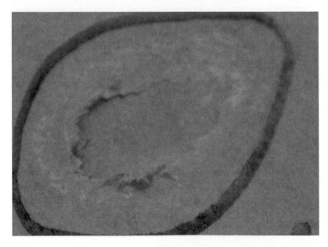

Figure A2.16 Surface erosion.

Excessive ply overlap: Occurs when the ply is not correctly trimmed during assembly. This can result in laminate dimensional tolerance errors. These errors could cause warping or induce high peel stresses.

Fastener holes: There is a wide range of fastener hole defect types. Figures A2.17, A2.18, and A2.19 show examples of typical fastener hole damage. The various fastener hole defects are discussed in the following list:

Fastener removal and reinstallation: Reworking a hole through fastener removal and reinstallation can result in local ply damage. Tensile strength appears to be insensitive and compression strength slightly sensitive to fastener removal and reinstallation.

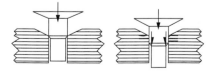

Figure A2.17   Drilled-hole exit damage.

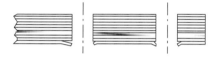

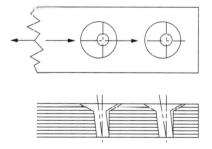

Figure A2.18   Tilted countersink fasteners.

Figure A2.19   Interference-fit fasteners.

Hole elongation: Overloading or bearing failure will result in elongation of the hole. The bearing/bypass load response will be affected. However, there appears to be little sensitivity to out-of-round holes.

Hole wear: Movement of the fastener in the hole will result in hole wear. The most aggressive wear will occur when the fastener shank pulls through the hole.

Improper fastener installation and seating: If the fastener is either over- or undersized and/or under- or overtorqued, the joint efficiency will be affected through changes to the bearing/bypass load response, and there is likely a substantial reduction in the bearing strength.

Missing fasteners: If a through-hole remains in the component when the fastener is missing, this will produce a stress concentration (Figure A2.20). The joint efficiency will also be affected. See **holes**.

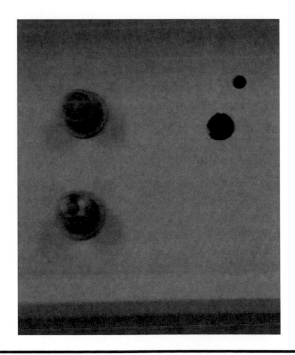

Figure A2.20   Missing fasteners.

Overtorqued fasteners: Result in local crushing of the outer plies (Figure A2.21). See **crushing**. Pull-through of the fastener can also result.

Hole exit-side damage: Caused by high drill-bit feed, which will produce **delaminations** on the back surface of the laminate (Figure A2.22). The criticality of the damage depends on the severity of the delaminations.

Other fastener hole problems: Other defect types that can occur in a fastener hole include: incorrect installation of interference-fit fasteners; pull-through; resin-starved bearing surface; and titled countersink holes. These are also sensitive in both tension and compression on the bearing face.

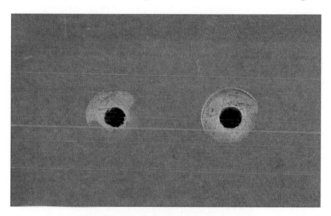

Figure A2.21   Surface damage due to overtorqued fasteners.

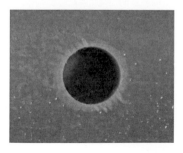

Figure A2.22   Drilled-hole exit damage.

Fiber distribution variance: Unevenness of fiber distribution or improper yarn spacing could change the laminate properties to the extent that the laminate load response will be different from design requirements. The effects will depend on the degree of variance of the fiber distribution, which is controlled by the curing process and is thus a manufacture-type defect (see Figure A2.23).

Fiber faults: See **damaged filaments** and Figure A2.23.

Fiber kinks: Sharp edge buckling of fibers within the matrix. Previous studies have concluded that kinking is a direct consequence of microbuckling. Excessive kinking of the fibers will eventually lead to fiber fracture, as shown in the micrograph of Figure A2.24.

Fiber/matrix debonds: Separation at the fiber/matrix interface (Figure A2.25). This will result in loss of shear transfer and degradation of the overall strength of the laminate. Fiber/matrix interface debonding results from excessive local shear-transfer stresses, particularly where short fibers are present. This type of defect is **matrix cracking** at the microscopic level.

Fiber misalignment: Occurs when there is either misorientation of the ply, deviation from predetermined

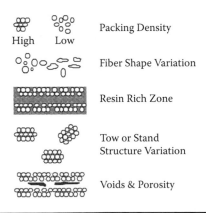

Packing Density
High   Low

Fiber Shape Variation

Resin Rich Zone

Tow or Stand
Structure Variation

Voids & Porosity

Figure A2.23   Microscopic defect types.

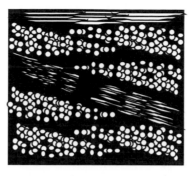

Figure A2.24   Fiber kinking and fracture.

Figure A2.25   Fiber-matrix debonding.

winding patterns, or washout of fiber from excessive
resin flow (Figure A2.26). Fiber misalignment is a fabrication control error. The effect is a localized change in
the load response of the laminate. Local misalignment
such as **fiber kinks** can cause fiber damage leading to loss of tensile strength or, under compressive
loading, can precipitate fiber buckling and premature
failure.

Fracture: Any type of cracking in the fibers, such as **fiber
kinks** as shown in Figure A2.25. Severe **matrix
cracking** is often associated with fracture. Fracture
can be the result of cuts in or penetration of the
laminate. Low-energy impact damage can also
result in back-surface fracture of thin laminates.
Impact-related fracture is illustrated in Figure A2.13.

Figure A2.26   Fiber misalignment.

Fiber fracture severely reduces the tensile strength of a laminate.

Holes and penetration: A result of either a design requirement, penetration damage, or repair action. Holes are generally a through-the-thickness defect, but partial through-the-thickness damage where there are fiber breaks can also be represented as a hole. Figures A2.15 and A2.27 depict holes and penetration damage. A hole in a laminate is a stress concentrator under any stress state. The severity of the stress concentrator will depend on the shape, condition, and location of the hole. Factors that will affect the hole severity include: drill burns, hole elongation, drill hole exit damage, miss-drilled holes, tilted holes, resin filled holes and free edge porosity. The type of penetration damage may include: battle damage, severe mishandling, countersink tear-out, or fastener pull-through. On a comparable size of penetration, strength losses are greater in holes than for BVID (barely visible impact damage).

Impact damage: The principle cause of penetration and the amount of damage depends on the energy level of the projectile involved (Figure A2.28). The amount of damage is also dependent on material properties, geometry, and the velocity of imparter and superimposed static loads. Damage propagation resulting from impact loads depends on loading type and strain levels, where

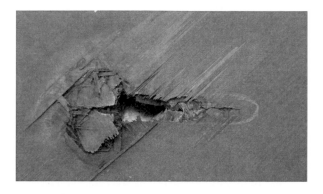

Figure A2.27   Projectile penetration of composite panel.

Figure A2.28   Impact damage of a composite skin sandwich panel.

the resulting strength losses can be conservatively
approximated on the basis of an "equivalent" round hole.

Marcelled fibers: Marcelling is waviness of the fibers
(Figure A2.29). Waves in the fibers will degrade the
lamina compression strength due to a decrease in
microbuckling resilience. Marcelled fibers are a result of
poor layup control.

Matrix cracking: See **cracks**.

Figure A2.29   Fiber marcelling.

Matrix crazing: Multiple cracks in all directions within the
resin. Nonstructural matrix materials are more prone to
this type of defect. Crazing is usually caused by over-
aged material or exposure to excessive heat and ultraviolet radiation. Matrix crazing is not applicable to epoxy
resin composites. See **cracks**.

Miscollination: A lack of straightness of the fibers and
somewhat similar to **marcelled fibers** (Figure A2.29).
Although not the same in principle as **fiber misalignment**, the effects are the same.

Mismatched parts: Mismatched parts are a tolerance error.
They will affect the load response of the laminate and
may cause fitting errors.

Missing plies: An incorrect stacking sequence is the consequence of missing plies. An asymmetric laminate may
be produced, from which out-of-plane and bending
stresses are induced. Also, the component design stress
distribution will be incorrect.

Moisture pickup: Moisture is absorbed into the laminate
through the matrix. Moisture contamination is usually contained in the outer plies, where degradation of
the resin properties such as softening will reduce the
stiffness and the fiber/matrix interface bond strength.
Swelling may also occur.

Nonuniform agglomeration of hardener agents: The
presence of a foreign body or resin-starved area in

the matrix. There will be a degradation of the local matrix properties.

Over-aged prepreg: A situation where the B-staged prepreg resin has aged or partially set to a point where the final cure will not provide adequate fiber/matrix adhesion and volatile evacuation. Reduction in the strength or stiffness of the laminate will result. Voids and porosity are also common in over-aged prepreg because the volatiles remain trapped within the resin.

Over/under cure: Occurs when the curing process is too long or too short in time and/or when too high or low a temperature is used. The laminate will have inadequate strength due to poor shear transfer in the fiber/matrix interface and poor stiffness response. This is a matrix defect.

Pills or fuzz balls: These are prepreg deviations, also known as "furring" of the fibers (Figure A2.30). When contained within a laminate, it can represent a contamination, where the area could also become susceptible to localized high peel stresses.

Ply underlap or gap: Occurs when the ply size is too short. This will cause localized inadequate strength and stiffness load response of the component as opposed to the design requirements.

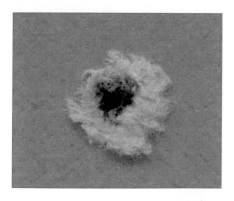

Figure A2.30   Fiber fuzzing of aramid after drilling a hole.

Porosity: Evidenced by the presence of numerous bubbles (voids) within the laminate, as illustrated in Figure A2.23. These bubbles result from: poor material or process control, over-aged material, moisture in the prepreg, or an autoclave malfunction. Although the bubbles are usually very small in size and randomly distributed, they produce localized stress concentrations. The size of the porosity is not as important as their concentration. Porosity degrades the tension, compression, interlaminar, and bearing properties of composites, but especially compression properties at elevated temperatures. The effect on the strength and fatigue lives of matrix-dominated laminates is more apparent. Very severe porosity levels of approximately 2% affect matrix-dominated properties significantly.

Prepreg variability: Caused by the exceedance of preset material property levels prior to cure. See **over-aged prepreg** for the effects on material load response.

Reworked areas: Areas of repair, either resin filled or patched, can result in localized strength and stiffness losses. The severity of these losses will depend on the repair type and the applied load level. Reworked area errors (Figure A2.31) can also contribute to degradation of structural performance.

Sandwich panels: Composite facings on sandwich panels can also be damaged or defective. Figure A2.32 illustrates the six principal forms of damage and defects in composite facings. Each can be referred to similar types of defects and damage as discussed in this Appendix.

Surface damage: Notches or any other surface irregularity resulting from mishandling or poor release procedures are termed as surface damage (Figure A2.33). For a further description, see **cuts and scratches**.

Surface oxidation: Can result from lightning strikes (Figure A2.34, local overheat) or battle damage (Figure A2.35, laser). The effect on structural integrity

Figure A2.31   Reworked area error (backing ply left in the repair scheme stack).

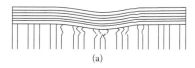

(a)

Figure A2.32a   Sandwich component damage types: core crushing.

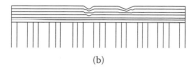

(b)

Figure A2.32b   Sandwich component damage types: face dimpling.

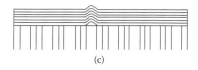

(c)

Figure A2.32c   Sandwich component damage types: laminate local instability (cell buckling).

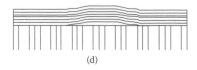

(d)

Figure A2.32d   Sandwich component damage types: separation from core (debond).

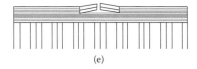

(e)

Figure A2.32e   Sandwich component damage types: filament and/or matrix fracture.

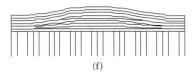

(f)

Figure A2.32f   Sandwich structure composite facing damage types.

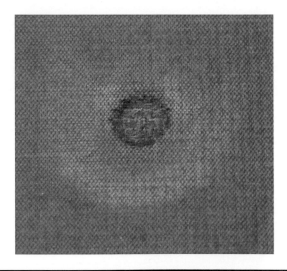

Figure A2.33   Surface burning (over-heated) damage.

Figure A2.34   Composite surface oxidation.

Figure A2.35   Composite surface swelling-blistering.

will be a degradation of the matrix properties. Surface oxidation is similar to other surface-type damage.

Surface swelling: Blisters caused by the use of undesirable solvents on the outer ply are examples of surface swelling (Figure A2.35). A localized breakdown of the matrix occurs, where a loss of fiber/matrix shear transfer and stiffness will result.

Thermal stresses: Although thermal stresses are not a true defect as such, the resulting residual stresses in the laminate are extraneous to the component's design and thus may affect the component's structural performance. Thermal stresses are a result of the curing process. Severe thermal stresses are known to occur in components upon removal from the autoclave or during cooldown in repairs. Cooldown rates must be carefully set and monitored.

Translaminar cracks: Through-the-thickness cracks where fibers are broken are translaminar cracks (Figure A2.7). Translaminar cracks are a direct result of extreme overload or impact damage and have the same effect as a hole stress concentrator. See **fracture** and **holes and penetration** for a further description.

Unbond or debond: Debonds are separations in a secondary adhesive bond or sandwich facing. An illustration of debonding is shown in Figure A2.32. They occur due to poor process control or fitting (Figure A2.36) and often by the inclusion of release film. In-service debonds are caused by impact damage, thermal spikes, overload, or freeze/thaw cycle. Debonds will reduce the local stiffness of a component; however, strength is not

Figure A2.36   Skin-to-core debond due to poor fitting.

necessarily affected. Basically, debonds are **delaminations** between bonded structures.

Variation in density: Variation in the density of the laminate is associated with resin inconsistencies, **voids**, or **porosity**. Their effect on the laminate is similar to that stated in **prepreg variability**.

Variation in fiber-volume ratio: Resin-rich or resin-starved areas produce variations in the fiber-to-matrix ratio (fiber-volume ratio). The fiber-volume ratio is an important parameter for determining the strength and stiffness of a laminate by micromechanics. Variation in the fiber-volume ratio is brought about by changes in the prepreg resin content through improper resin bleed-out during cure. For further information and description, see **prepreg variability** and **over/under cured**. Figure A2.23 illustrates variations in the resin fraction.

Variation in thickness: Normally associated with inconsistencies in the resin content of the laminate or the adhesive layer in a bonded joint (Figure A2.37). As a result, excessive peel stresses may occur, particularly in a bonded joint. The bond joint efficiency may also be degraded.

Voids: Voids are trapped air or other volatiles in the resin (Figure A2.38). They are caused by poor process control and can be localized or uniformly distributed (porosity).

Figure A2.37   Adhesive layer thickness variation during assembly.

Figure A2.38   Voids in resin filler.

Voids are a single bubble, whereas **porosity** is a cluster of several microscopic voids. Voids generate from:

1. Dissolved air within the resin
2. Air stirred into the resin
3. Trapped air in a filament bundle
4. Residual solvent carrier
5. Reaction products from the curing process
6. Volatilization of low-molecular-weight components of the resin, or of organic inclusions, at high cure temperatures

The extent to which voids produce deterioration in mechanical and other properties is a function of void content, void distribution, and void shape. Voids reduce the magnitude of the following mechanical properties: interlaminar (short beam) shear strength, longitudinal and transverse flexural and tensile strength and modulus, compressive strength and modulus, and fatigue resistance. However, only rather large voids reduce interlaminar shear strength significantly.

Warping: Warping is a result of detailed or assembly part mismatch. Also, residual thermal stresses remaining in the laminate after fabrication can produce warping (Figure A2.39). The main concerns are that the part may not fit at the next assembly or, if fitted, out-of-plane stresses may be induced.

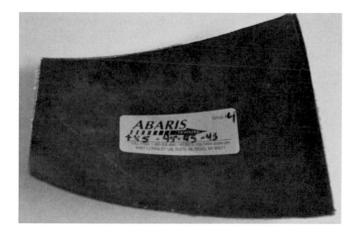

Figure A2.39   Thermally warped panel due to layup error.

Wrong materials: Wrong materials used in the fabrication of the component are a blueprint error. The resulting component's stress and stiffness characteristics will not match that of the design requirements and therefore may produce an inferior component.

Chapter 3

Finding the Nonconformity

Introduction

Nondestructive inspection (NDI) methods are employed in the repair process of composite and bonded structures in three ways:

1. Damage location
2. Damage evaluation, i.e., type, size, shape, and internal position
3. Post-repair quality assurance

The first and most important activity in a repair process is to identify the defect or damage. Assessment of the damage is initially achieved by visual inspection. This localizes the damaged area, and then a more sensitive NDI method is employed to map the extent of any internal damage. Detailed NDI is very important when dealing with composite and bonded structures because damage is often hidden within the structure, with little to no surface indication. See Figures 3.1 and 3.2 for respective examples of visible and nonvisible composite damage.

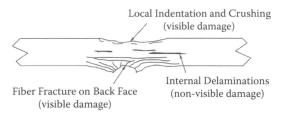

Local Indentation and Crushing
(visible damage)

Fiber Fracture on Back Face
(visible damage)

Internal Delaminations
(non-visible damage)

Figure 3.1   Internally hidden damage with external visible damage.

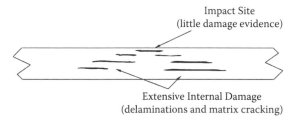

Impact Site
(little damage evidence)

Extensive Internal Damage
(delaminations and matrix cracking)

Figure 3.2   Barely visible impact damage (BVID).

The types of NDI methods currently available are:

■ Visual, which includes methods with optical magnification and defect enhancement
■ Acoustic methods, which identify changes in sound emission
■ Ultrasonic methods, such as A-scan and C-scan
■ Thermography
■ Interferometry
■ Radiography
■ Microwave
■ Material property changes, i.e., stiffness and dielectric

Each of these methods is briefly discussed in the following sections. A more detailed examination of the various NDI methods can be found in references such as ASM International Engineered Materials handbooks (1988, 1990), Hoskins and Baker (1986), Summerscales (1987), and Hsu (2008), or see the Bibliography at the end of this book.

NDI Methods

The majority of NDI methods available for use on composite and bonded structures have been used successfully with metallic structures. However, for application of these NDI methods on composite and bonded structures, some changes to operating parameters and results interpretation are required. Due to the diversity of defect types likely to be found in composite structures, several methods may be required to fully detail the damage state. Thus, with several methods required to find and assess the damage type in composite structures, a larger investment in NDI equipment and correspondingly more highly trained NDI assessors are needed.

Visual Inspection

Apart from simply using the assessor's eye, which only identifies obvious defects such as that shown in Figure 3.2, simple magnification can identify quite small surface defects. To improve the visual clarity of defects or matrix cracks, enhancement with a dye penetrant can be used. Some internal defects can also be found using boroscope methods, but access is still required, i.e., through a fastener hole. Bondline visual inspection will provide some assessment of the resin flow. The typical resin flows from a bonded joint edge, shown in Figure 3.3, can be judged by simple visual means.

Visual methods can be summarized as follows:

- ■ Visual methods:
    - Are inexpensive
    - Are simple
    - Require low skill levels
    - Need the surface in question to be relatively clean
    - Are suitable for surface defects only

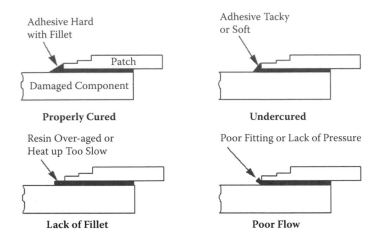

Figure 3.3   Bonded-joint-edge resin flows.

■ Dye-penetrant methods:
- Contaminate the surface to be inspected
- Are suitable for surface defects only
- Require pre-cleaning and post-cleaning of the part
- Are portable
- Are simple to apply
■ Require some level of operator skill

Acoustic Methods

All of the acoustic emission methods currently considered for composite structures require the operator or acoustic noise-detecting equipment to listen to crack growth and movement via changes in sound from a light impact or propagation of elastic wave energy. Acoustic methods include:

■ Coin-tap (hammer) method (Figure 3.4):
- Is simple to use
- Is suitable only for near-surface defects (shallow)
- Is geometry dependent

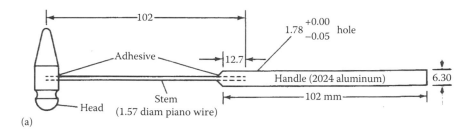

(a)

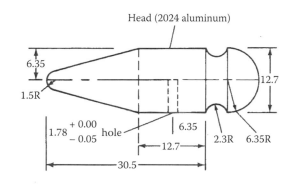

(b)

Figure 3.4   Tap-test hammer (dimensions in mm).

- Usually requires two-sided access for sandwich panels
  Is portable
- Requires the operator to have a good ear
■ Acoustic emission (Figure 3.5):
  - Requires experienced operators to set up the system
    and interpret the results
  - Has complex output to be understood
  - Is portable and recordable
■ Is reasonably sensitive to small changes in defects

Ultrasonic Methods

The implementation of ultrasonic inspection can range
from inexpensive to quite costly in terms of the equipment
required to undertake the process. The methods can simply
provide details of depth and size of the nonconformity, or

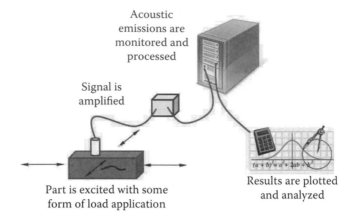

Acoustic emissions are monitored and processed

Signal is amplified

Part is excited with some form of load application

$(a + b)^2 = a^2 + 2ab + b^2$

Results are plotted and analyzed

Figure 3.5   Acoustic emission detection of corrosion in honeycomb sandwich panels.

full details of the topography of subsurface defects. There are two principal ultrasonic methods: pulse-echo (A-scan) or through transmission (C-scan). Both methods measure change in sound attenuation (amplitude loss) as the sound passes through the area of interest. C-scan ultrasonic methods are illustrated in Figures 3.6–3.8, and the typical C-scan results are shown in Figure 3.9. The types of C-scan ultrasonic NDI methods are:

- Pulse-echo method:
    - Displays amplitude of the return signal versus time
    - Requires a coupling agent to allow sound wave through the transducer to the part
    - Provides information on defect type, size, location, and depth
    - Is reasonably sensitive to find (identify) small defects
    - Requires a standard specimen to compare the results with
    - Requires experienced operators
    - Is portable and recordable
    - Requires pre-cleaning and post-cleaning of the part

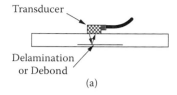

(a)

Figure 3.6a   Ultrasonic inspection techniques: contact pulse-echo.

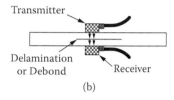

(b)

Figure 3.6b   Ultrasonic inspection techniques: contact through transmission.

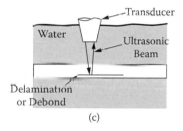

(c)

Figure 3.6c   Ultrasonic inspection techniques: immersion pulse-echo.

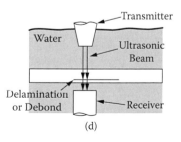

(d)

Figure 3.6d   Ultrasonic inspection techniques: immersion through transmission.

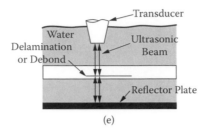

(e)

Figure 3.6e   Ultrasonic inspection techniques: immersion reflection.

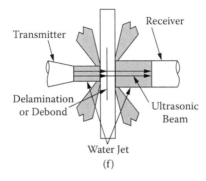

(f)

Figure 3.6f   Ultrasonic inspection techniques: water jet through transmission.

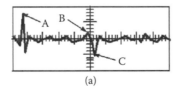

(a)

Figure 3.7a   Representative ultrasonic pulse-echo results of a graphite/ epoxy composite skin and honeycomb core: a well-bonded sample.

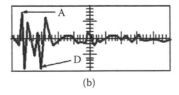

(b)

Figure 3.7b   Representative ultrasonic pulse-echo results of a graphite/epoxy composite skin and honeycomb top: skin delamination.

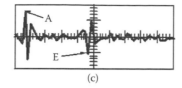

(c)

Figure 3.7c   Representative ultrasonic pulse-echo results of a graphite/epoxy composite skin and honeycomb: debond between skin and core.

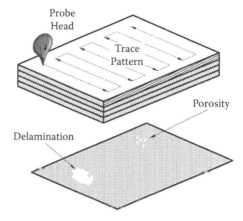

Figure 3.0   Schematic of the ultrasonic C-scan.

Figure 3.9 (See color insert.)   Ultrasonic C-scan results of a graphite/ epoxy laminate.

■ Through transmission (immersion) method, which is basically the same as pulse-echo, except that it:
  - Is automated and therefore faster
  - Only provides accurate definitions of defect size and location

     – Provides full coverage of the component
     – Requires double-sided access
■ Is for internal defects only

Thermography

Thermography is an NDI technique that measures the response of a structure to either thermal energy dissipation or induced temperature through thermoplastic characteristics. Passive thermography methods identify internal noncontacting defects, where the rate at which thermal energy dissipates is reduced. Active thermography activity vibrates or load-cycles the structure, and the localized stress increase in the presence of a defect generates heat. Both thermography methods:

■ Require standards to verify results
■ Are portable and recordable
■ Require experienced operators to operate and assessors to interpret
■ Are geometry dependent

    Typical thermography results are shown in Figure 3.10.

Interferometry

The use of light and its reflective properties to identify defects is known as *optical interferometry.* There are three basic methods: moiré (including shadow moiré),

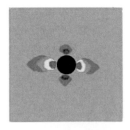

Figure 3.10 (See color insert.)   **Schematic of thermography results.**

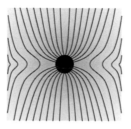

Figure 3.11 Schematic of moiré interferometry of composite laminate with hole.

holography, and shearography (see Figures 3.11–3.13). Interferometric methods:

- Typically require expensive equipment
- Need skilled equipment operators and interpreters
- Are generally not portable
- Provide a full-field record of the defect behavior under load
- Are very sensitive
- Show how the structure and defect react under loading

Radiography

The principle of radiography as an NDI technique is illustrated in Figure 3.14. The two main radiography methods are X-ray and neutron radiography. Typical X-ray radiography results are shown in Figures 3.15 and 3.16.

- The X-ray method:
    - Is easy to interpret
    - Is excellent for honeycomb sandwich panel inspection
    - Provides a permanent record
    - Requires expensive equipment
    - Can be portable
    - Requires strict safety procedures
    - Requires experienced operators
    - Can be enhanced with dye penetrants

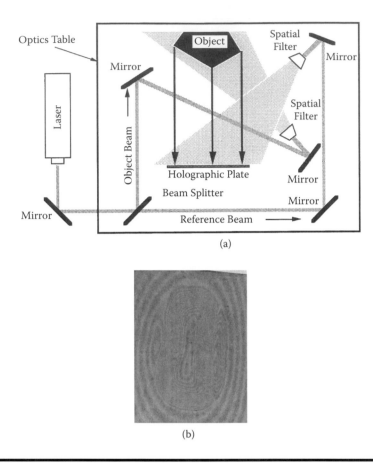

(a)

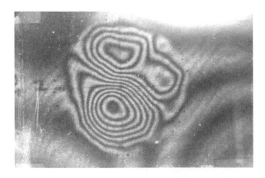

(b)

Figure 3.12   Double-exposed holography showing defects.

Figure 3.13   Defect evaluation using shearography.

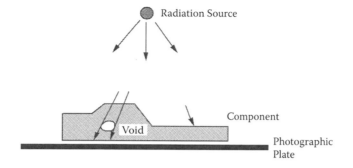

Figure 3.14   Principle of radiography.

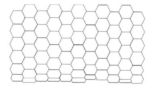

Figure 3.15   Schematic of X-ray radiographs of honeycomb core: edge crush.

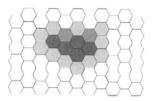

Figure 3.16   Schematic of X-ray radiograph of honeycomb core with water entrapment.

- ■ Neutron radiography:
  - – Is very expensive to use
  - – Is basically similar to the X-ray method
  - – Is excellent for composites and moisture entrapment
- ■ Provides better resolution of results

Microwave

Microwave NDI is used mainly on *nonmetallic* materials to determine the degree of moisture content through the measurement of microwave absorption. The method requires:

- Two-sided access to the composite panel or structure
- Application of metal component shielding
- Significant operator safety
- The component to be clean

Material Property Changes

The two major NDI methods based on material property changes are *dielectric* and *stiffness*. Dielectric measures the degree of adhesive or composite cure by reading inductance, and stiffness changes are determined from mechanical testing and comparison with theoretical analysis. Both methods have their limitations.

In Situ *Methods*

Several *in situ* methods are currently explored in composite laminated structure. These *in situ* methods are briefly discussed here:

- *Optical fibers*: The embedding of optical fibers into the laminated structure provides a source of localizing damage only if the optical fiber is also damaged. However, with the optical fiber damaged, the precise location of the damage can be quickly identified. This is ideal for BVID (barely visible impact damage) or NVID (nonvisible impact damage) in laminated materials. The optical fiber is typically placed close to the surface of the structure that is likely to be damaged and spaced appropriately to ensure a high reliability of damage detection.
- *Piezoelectric sensors*: Piezoelectric sensors provide an electrical current when subjected to stress or

deflection. These electrical currents can identify local overload conditions and, in particular, out-of-plane loadings. Like optical fibers, piezoelectric sensors are placed in the areas of greatest concern or likelihood of damage.

■ *Embedded strain gauges*: Along with the same reasoning as piezoelectric sensors, the embedded strain gauge will identify local overload conditions through the generation of excess strain readings. Placement of the embedded strain gauges is very important, and the interpretation of the strain data is crucial in obtaining relevant overload outcomes.

A significant issue with the use of embedded sensors is the repair of the sensor if the composite is damaged.

Application of NDI Methods

Successful application of any NDI method depends on the selection of the most suitable method as well as the availability of personnel with the required skills.

NDI Selection Process

The NDI selection process is based on:

■ The configuration of the component and the materials it is made from
■ The type and size of defects to be inspected
■ Accessibility to the assessment area
■ The availability of both equipment and skilled operators

The ability of the various NDI methods to find various defects in composite and bonded structures is listed in Table 3.1.

Table 3.1 NDI Methods versus Defect Type

NDI Method	Defect	Visual	Penetrant	Tap	Bond Tester	Pulse/Echo	Through Transmission	X-Ray	Dielectric	Thermography	Interferometry	Microwave	Neutron Radiography	Mechanical Impedance
Laminate	Delaminations	1,2	1	✓	✓	✓	✓	3		✓	✓			✓
	Macrocracks	1,2	2	✓	✓			3		✓	✓			
	Fiber fracture							✓		2,3	2,3			
	Interfacial cracks									2,3	✓			
	Microcracks	1	1	2	2					✓	✓			
	Porosity	1		2	2	✓	✓	✓		2	✓			✓
	Inclusions	1		2	2	2	2	✓		✓	✓			
	Heat damage	1		2	2				2		2			
	Moisture							2	✓		2	✓	✓	

Voids			2		✓	✓	✓	✓	✓	✓	✓		
Surface protrusions	✓					✓	✓				✓		
Wrinkles	✓				✓	✓							
Improper cure								✓	2	2		✓	
Bondline													
Debonds	1,2	1	✓	✓	✓	✓	✓	✓	✓	✓	✓	✓	
Weak bonds								2	2	✓			
Cracks	1,2	1	2	2	2	3		✓	✓	✓			
Voids			✓	✓	✓	✓	✓	✓	✓	✓	✓		
Moisture					✓	✓	✓	✓		2	✓	✓	
Inclusions			2	2	2	✓	✓	✓	✓	✓	✓		
Porosity			2	✓	✓	✓			✓	✓			
Lack of adhesive			✓	✓	✓	✓	✓	✓	✓	✓	✓		
Sandwich panels													
Blown core			✓	✓	✓	✓						✓	✓
Condensed core			2	2	2	✓						✓	✓
Crushed cure			2	2	2	✓						✓	✓

(Continued)

Table 3.1 (Continued) NDI Methods versus Defect Type

NDI Method → / Defect ↓	Visual	Penetrant	Tap	Bond Tester	Pulse/Echo	Through Transmission	X-Ray	Dielectric	Thermography	Interferometry	Microwave	Neutron Radiography	Mechanical Impedance
Distorted core			✓				✓					✓	
Cut core				✓		✓	✓					✓	
Missing core			2	2	2	2	✓					✓	✓
Node debond							✓		2	✓			
Water in core			2	2		2	✓				✓	✓	
Debonds			✓	✓	✓	✓	✓		✓	✓			✓
Voids			2	2	✓	✓	✓		✓	✓			
Core filler cracks			2	2	2	✓	3		2	2			
Lack of filler			2	2	2	✓	✓		2	2		✓	

Note: 1 = open to surface; 2 = unreliable detection; 3 = orientation dependent.

NDI Personnel

The NDI operator and assessor must be:

- Conversant with several different inspection techniques
- Able to set up the equipment and effectively modify the standard diagnostic arrangements to suit the target
- Skilled to interpret the resulting NDI information
- Knowledgeable of safety standards and procedures
- Able to comply with MIL-STD-410 or its equivalent

Important Requirements

For NDI to be successful in detecting the extent of damage in composite and adhesively bonded structures and components, three requirements must be satisfied. They are as follows:

1. *Equipment and facilities*: The suitable NDI equipment and facilities, including personnel safety and environmental health procedures, must be available, calibrated, and in good working order.
2. *Trained operators*: The operators of NDI equipment must be adequately trained and experienced to ensure that the results from any damage assessment survey are both accurate and reliable.
3. *Comparative specimens*: Any NDI technique is comparative in nature, i.e., the results of an assessment survey are usually compared with a good or a like damaged specimen. This is particularly important when calibrating NDI equipment.

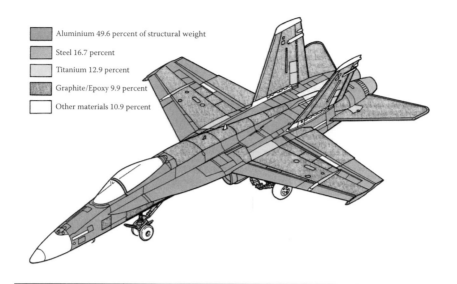

Aluminium 49.6 percent of structural weight

Steel 16.7 percent

Titanium 12.9 percent

Graphite/Epoxy 9.9 percent

Other materials 10.9 percent

Figure 1.2   Material breakdown of the F-18 Hornet airframe (United States Navy, 1984).

Figure 1.6   Wright Military Flyer B Model (1909).

Figure 1.7   Supermarine Spitfire (1940).

Figure 1.8   de Havilland DH98 Mosquito (1938).

Figure 1.9   Lockheed P-2V Neptune (1947).

Figure 1.10   Boeing/MDD Harrier.

Figure 1.11   Boeing/Bell V-22 Osprey.

Figure 1.12   Airbus A380.

Figure 1.13   Boeing/MDD C-17 Globemaster III.

Figure 1.14   Sports aircraft.

Figure 1.15   X-34 (NASA).

Figure 1.16   Wind turbines.

Figure 1.17   Luxury cruiser.

Figure 1.18   Auto racing.

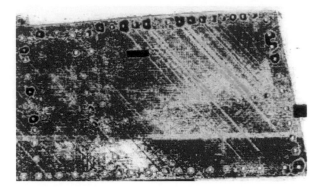

Figure 3.9   Ultrasonic C-scan results of a graphite/epoxy laminate.

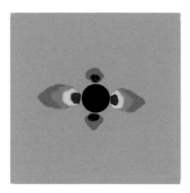

Figure 3.10   Schematic of thermography results.

Chapter 4

Failure Mechanisms

Introduction

Knowledge of the failure modes and mechanisms in composite materials and structures will provide information enabling the most appropriate stress analysis method to be used in determining defect and damage criticality. The possible failure modes can range from simple loss of structural stiffness due to instantaneous first ply failure, through to reduction in load-carrying capacity due to localized deformation and damage growth, or complete loss of load-carrying capacity because of the presence of the defect or damage. This chapter first reviews the various failure modes and mechanisms associated with composite materials and structures, then details both the simple and complex failure modes of composite materials, and finally discusses the mechanisms of fracture and failure in the composite material that leads to ultimate loss of performance.

Failure Modes and Mechanisms

The level of structural degradation in engineering material properties varies with several factors. These factors are defined as follows:

Defect severity: The severity of the defect will be determined from the stress state of the defect on the local composite structure. The stress state will be briefly discussed in Chapter 5.

Defect location and orientation: Specific locations of the defect will have a different impact on the defect stress severity. The physical orientation of the defect can result in the defect being either severe or benign.

Frequency of defect occurrence: The frequency of the defect's occurrence will also affect the severity or impact on the structural integrity of the composite material.

Component load path criticality and stress state: The load state will interact with the defect location and orientation to either increase or reduce the defect criticality.

Defect idealization: Previously, the defect type was generalized to be either intralaminar matrix cracks, interlaminar matrix cracks, or fiber fraction or design variance. This idealization may not be adequate for the stress analysis and determination of defect criticality.

Design load levels and nature: The loading levels in the component and the nature of the load (static or dynamic) will also impact the severity of the stress state in the presence of the defect.

Defect detectability and detection capabilities: The ability to find the defect and then classify the defect type will depend on the capabilities of the detection equipment and the skill level of the operator. This is a very crucial aspect of defect criticality identification.

Local repair capabilities: In addition to defect detectability, another important consideration is the repair capabilities of the facility involved in restoration of composite performance. The local repair capabilities may not be able to rework the damaged or defective area back to operative levels. Also, the repair capability may increase the loss of structural integrity if skill and knowledge are inadequate.

Component configuration: The configuration of the component leads to the specific types of defects and damage while also influencing how the defect reacts to imposed loading. Careful examination of the component configuration is required to ensure full understanding of its impact on the presence of the defect.

Environmental condition: The environmental condition of the composite component will typically be based on the moisture content of the composite and the operational temperature. Other environmental conditions, such as seawater and other fluids, need to be considered and understood as to their impact on the structural performance with defects.

Loading history: The load history may impact how the defect influences the changes in structural integrity. Loading conditions on the defect prior to the defect's identification may provide an indication of the structural defect criticality.

Material property variations: The variation in material properties, in particular fiber-volume ratio, can have a significant impact on the structural performance in the presence of defects and damage. This is particularly true for defects in the matrix and matrix-dominated behavior.

Acoustic vibration response: High-frequency cyclic loading is of particular interest for several defect types. Thus acoustic vibration is considered a specific focus in the understanding of defect failure modes and response.

Basic Modes of Failure

There are four types of failure modes distinguishable in composite materials. Each of the four types of failure can act independently, or as a pair, or all together. The four failure modes are defined as:

1. *Fiber failure*: Any particular fiber breakage is considered a fiber failure. However, buckling fibers can also be a form of fiber failure, since the fiber is unable to support the design load.
2. *Transverse matrix failure*: The failure of the matrix within a ply is considered as the transverse failure or cracking of the resin system and is independent of the fibers. Transverse matrix cracking is also known as intralaminar cracking. Matrix property degradation due to environmental aggravation (chemicals, fluids, or moisture); and aging under solar radiation, thermal cycling and attack; and other forms of radiation—all of which attack and degrade the resin properties—also can be considered transverse matrix failure.
3. *Interfacial failure*: Failure of the bondline interface between fiber and matrix is the interfacial failure of composite materials. This failure mode is primarily attributed to poor resin/fiber selection, but can occur with aging due to environmental conditions.
4. *Delaminations*: Failure of the interface between adjacent plies where separation occurs is termed as a delamination. Delaminations are also known as interlaminar cracking. Delaminations are another form of matrix cracking, but the failure locus is within the plane of the local laminated structure.

All of the basic failure modes are distinguished by their macroscopic failure characteristics, which consist of microscopic-initiated cracking in the matrix or fiber.

Under in-plane loading, the failure appearance can be described as follows:

Longitudinal tensile failure: Longitudinal tensile failure shown in Figure 4.1 is characterized by either brittle fiber fracture (Figure 4.1a), brittle fiber fracture with fiber pullout (Figure 4.1b), or staggered failure (Figure 4.1c). Staggered failure is brittle fiber fracture and fiber pullout combined with matrix shear and longitudinal splitting.

Longitudinal compression failure: Longitudinal compression failure consists of fiber microbuckling, matrix yielding, panel buckling, shear failure, or ply delamination by transverse tension. These failure modes are illustrated in Figures 4.2, 4.3, and 4.4.

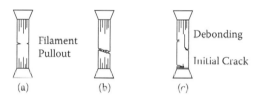

Figure 4.1   Longitudinal tensile failure modes.

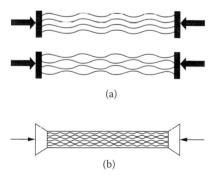

Figure 4.2   Longitudinal compression failure mode: microbuckling; (a) asymmetric buckling, (b) symmetric buckling.

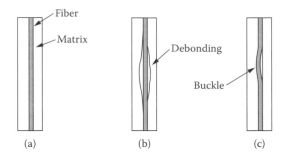

Figure 4.3 Longitudinal compression failure mode: constituent debonding.

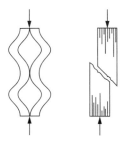

Figure 4.4 Longitudinal compression failure mode: panel buckling-splitting and shear.

Transverse tension failure: Transverse tension failure can be matrix tension failure, fiber/matrix interface debonding, or longitudinal splitting of the fibers. Figure 4.5 depicts the matrix tension mode of failure. This is a preferred failure mode for transverse tension loading as it tends to be the weak link with appropriate selection of the resin.

Transverse compression failure: Transverse compression failure consists of matrix compression failure, matrix shear failure, fiber/matrix interface shear failure, or fiber crushing. Figure 4.6 illustrates the most common of these modes.

Intralaminar shear failure: Intralaminar shear failure covers matrix intralaminar shear, matrix shear, or matrix/fiber debonding (Figure 4.7).

Woven-cloth failure modes: Woven cloth will exhibit many of the failure modes previously discussed that pertain to

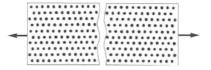

Figure 4.5   Transverse tension failure mode.

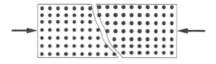

Figure 4.6   Transverse compression failure mode.

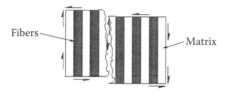

Figure 4.7   Intralaminar shear failure mode.

unidirectional composite plies. However, the cross-woven fibers (the fill fibers) have a major influence in inhibiting many of the matrix-dominated failure conditions. Woven-cloth failure modes consist of the following:

Fiber tensile fracture: Fiber fracture in woven-cloth composite plies is much the same as for unidirectional plies. The fracture path is very much localized over a specified perpendicular plane to the tensile loading direction, as illustrated in Figure 4.8.

Fiber compression microbuckling: The microbuckling of the warp fibers is more pronounced in woven composites due to the already kinked fiber tows/yards. The warp fiber is constrained from microbuckling by the through-the-thickness tensile strength of the matrix material that binds the warp fiber to the fill

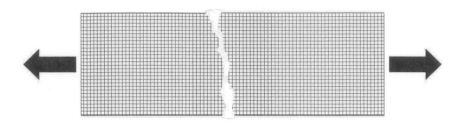

Figure 4.8   Tensile failure mode in woven-cloth plies.

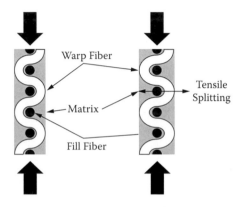

Figure 4.9   Microbuckling of woven-cloth composites.

fiber (Figure 4.9). Harness stains tend to microbuckle at lower loads than plain-weave cloth composites.

Matrix tensile fracture: Under tensile loads (axial or transverse), the matrix fractures slightly before the fibers fracture. First-ply failure (matrix cracking) is typically at 85%–95% of ultimate load (fiber fracture).

Fiber/matrix interfacial compression splitting: In combination with microbuckling, the splitting of the matrix/fiber bonds occurs prior to unstable fiber tow/yarn buckling (see Figure 4.9).

Matrix shear fracture: The fracture of the matrix in shear is most common with angle fiber cloth layups (Figure 4.10).

The failure modes of unidirectional plies, under simple loading conditions, are summarized in Table 4.1. A similar summary for woven cloth materials is provided in Table 4.2.

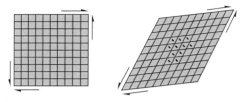

Figure 4.10   Matrix shear failure of woven angle-ply layups.

Complex Modes of Failure

Because of the laminated nature of fiber composite materials, the modes of failure can be quite complex. This complexity originates in that crack initiation resembles one of the typical modes detailed previously for a unidirectional laminate, but it propagates to another mode of failure and ultimately fails catastrophically by fiber failure.

Angled fiber composites: The simplest failure modes in the complex system are those of angled fiber composites. Angled fiber composites all show failure initiating by intralaminar cracking and then delaminations. The three possible failure modes in an angled-ply laminate are illustrated in Figure 4.11. These modes are controlled by the fiber orientation percentage in the laminate. The failure mode is predominantly an interlaminar shear initiated for small fiber angles and then propagating to fiber breakage (FB mode). This mode shifts to a combined interlaminar normal, in-plane shear, and in-plane normal fracture (fiber shear/delamination [FS/DEL]) for intermediate angles and is predominantly transverse tension for larger fiber angles.

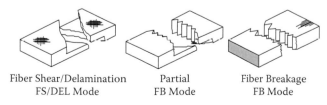

| Fiber Shear/Delamination | Partial | Fiber Breakage |
| FS/DEL Mode | FB Mode | FB Mode |

Figure 4.11   Three possible failure modes for the angled-ply laminate.

Table 4.1   Failure Modes of Unidirectional Plies

Mode of Failure	Nature of Loading	Primary Factors	Secondary Factors
1. Fiber failure transmitted laterally (brittle failure)	Longitudinal tension	(−) Fiber tensile strength (−) Fiber volume fraction (+) Matrix stiffness and strength (+) Interface bond strength	(−) Cure shrinkage stresses
2. Fiber failure transmitted longitudinally and laterally (brushing)		(−) Fiber tensile strength (−) Fiber volume fraction (−) Matrix stiffness and strength (−) Interface bond strength (−) Cure shrinkage stresses	(+) Fiber strength variability
3. Brittle fiber failure (inclined shear)	Longitudinal compression	(−) Fiber strength and volume fraction (+) Local and overall stability (−) Local distortion/ eccentricity	

Table 4.1 (*Continued*)   Failure Modes of Unidirectional Plies

Mode of Failure	Nature of Loading	Primary Factors	Secondary Factors
4. Kink-band failure (timberlike)		(−) Matrix shear stiffness (+) Local distortion/ eccentricity (−) Fiber diameter	(−) Fiber shear stiffness (−) Fiber volume fraction (+) Moisture and temperature (matrix)
5. Fiber microbuckling		(−) Matrix transverse stiffness (+) Fiber diameter/ eccentricity (−) Matrix tensile strength	(+) Moisture and temperature (matrix)
6. Locally oriented delamination		(+) Local distortion/ eccentricity (−) Fiber diameter and modulus (−) Matrix tensile strength	(−) Matrix shear stiffness (+) Moisture and temperature (matrix)

(*Continued*)

Table 4.1 (*Continued*)   Failure Modes of Unidirectional Plies

Mode of Failure	Nature of Loading	Primary Factors	Secondary Factors
7. Matrix/ bondline tension fracture	Transverse tension	(−) Matrix/ fiber average failing strain (−) Fiber/ matrix bond strength (+) Cure shrinkage stresses	(+) Displaying between fiber and matrix stiffness (−) Matrix strength (+) Fiber distribution irregularity (+) Fiber volume fraction
8. As for Mode 7 (delamination possible)	Short transverse tension (normal)	As for Mode 7 (+) Curved laminate bending	As for Mode 7
9. Matrix/ bondline (inclined shear fracture)	Transverse compression	As for Mode 7	As for Mode 7
10. Transverse layer buckling		(−) Matrix/fiber transverse modulus (−) Matrix/fiber shear modulus	(+) Fiber volume fraction
11. Shear in matrix, fiber/ matrix debonding, interlaminar shear in laminate	Longitudinal/ short transverse shear	(−) Matrix shear strength (−) Fiber/matrix adhesion (+) Moisture and temperature (matrix)	(+) Fiber volume fraction (+) Cure shrinkage stresses

Table 4.1 (*Continued*) Failure Modes of Unidirectional Plies

Mode of Failure	Nature of Loading	Primary Factors	Secondary Factors
12. Shear in matrix, interlaminar shear in laminate	Longitudinal/ transverse shear	(−) Matrix shear strength	(+) Cure shrinkage stresses (+) Moisture and temperature (matrix)
13. Shear in matrix, (cross-fiber shear) interlaminar shear in laminate	Transverse/ short transverse shear	As for Mode 10	As for Mode 10

Source: Revised from ESDU 82025.

(+): The greater the quantity, the more likely it is to fail.
(−): The lower the quantity, the more likely it is to fail.

Multidirectional laminates: The failure modes of multi-directional laminates incorporate the failure modes of on-axis and off-axis unidirectional laminates, angled fiber laminates, and matrix cracking failure modes. The ultimate failure is often very complex. Usually one of the previously mentioned failure modes will initiate at the individual-ply level and proceed to ultimate failure by another failure mechanism. An excellent summary of multidirectional laminate failure modes is provided in Table 4.3. The state of stress in a multidirectional laminate is truly three dimensional, and therefore the significance of Mode I and Mode III loadings are greater than for unidirectional laminates, where Mode I loading failures predominate. For a given matrix crack, its propagation is controlled by the opening (Mode 1), shearing (Mode 2), tearing (Mode 3), or mixed-mode failures (see Figure 4.12).

Table 4.2 Failure Modes of Woven Plies

Mode of Failure	Nature of Loading	Primary Factors	Secondary Factors
1. Fiber failure transmitted laterally (brittle failure)	Longitudinal tension	(−) Fiber tensile strength (−) Fiber volume fraction (+) Matrix stiffness and strength (+) Interface bond strength	(−) Cure shrinkage stresses
2. Fiber failure transmitted longitudinally and laterally (brushing)	Transverse tension	(−) Fiber tensile strength (−) Fiber volume fraction (+) Matrix stiffness and strength (+) Interface bond strength	(−) Cure shrinkage stresses
3. Fiber failure transmitted laterally (brittle failure)	Longitudinal compression	(−) Fiber tensile strength (−) Fiber volume fraction (+) Matrix stiffness and strength (+) Interface bond strength	(−) Cure shrinkage stresses
4. Fiber failure transmitted longitudinally and laterally (brushing)	Transverse compression	(−) Fiber tensile strength (−) Fiber volume fraction (+) Matrix stiffness and strength (+) Interface bond strength	(−) Cure shrinkage stresses

Table 4.2 (*Continued*)   Failure Modes of Woven Plies

Mode of Failure	Nature of Loading	Primary Factors	Secondary Factors
5. Fiber failure transmitted laterally (brittle failure)	In-plane shear	(−) Fiber tensile strength (−) Fiber volume fraction (+) Matrix stiffness and strength (+) Interface bond strength (−) Fiber compressive strength	(−) Cure shrinkage stresses

(+): The greater the quantity, the more likely it is to fail.
(−): The lower the quantity, the more likely it is to fail.

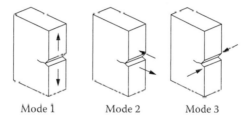

Mode 1   Mode 2   Mode 3

Figure 4.12   Three basic modes of failure.

Loading mechanisms: The loading mechanism in a composite laminate also governs the failure modes. Here the three principal loading methods are discussed briefly: tension, compression, and shear.

Tension: The fracture process of composites under tensile loading depends on the fiber angle. By characterizing failure surfaces and associated load angle ranges, the predominant failure modes of off-axis tensile specimens are:

0° fibers: Irregular fracture surface with fiber pullout

5° to 30°: Regular fracture surface with extensive matrix lacerations

Table 4.3   Failure Modes of Multidirectional Laminates

Mode of Failure	Nature of Loading	Primary Factors	Secondary Factors
Layer transverse tension cracking (regularly spaced crack through independent layers)	Tension, shear, and compression	(+) Transverse tensile strain components in layer, crack frequency increases with strain (+) Cure shrinkage stresses	Some effective transverse and stiffness retained initially via uncracked zones
Layer longitudinal tension fracture		As unidirectional longitudinal tensile (+) Stress concentration from adjacent cracked layers	Laminate layup and stacking sequence (+) Cure shrinkage stresses
Delamination/ layer free edges or notches		(−) Interlayer shear strength (+) Moisture and temperature (resin), varies with layup and stacking sequence	Tension or shear, delamination often local to raisers; usually catastrophic in compression
Delamination/ layer buckling		(+) Local layer distortion (+) Layer thickness (−) Longitudinal fiber modulus (−) Matrix tensile strength	(−) Matrix shear stiffness (+) Moisture and temperature (resin)

Table 4.3 (*Continued*)   Failure Modes of Multidirectional Laminates

Mode of Failure	Nature of Loading	Primary Factors	Secondary Factors
Interlaminar shear	Short transverse shear (normal)	As for Modes 11 and 13 (unidirectional), will vary with layer stacking sequence	As for Modes 11 and 13 (unidirectional)
Interlaminar tension	Normal tension	As for Modes 7 (unidirectional), will vary with layer stacking sequence	As for Modes 7 (unidirectional)

Source: ESDU 82025.

(+): The greater the quantity, the more likely it is to fail.
(−): The lower the quantity, the more likely it is to fail.

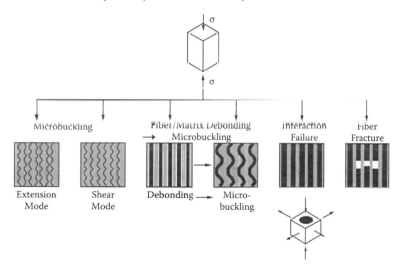

Figure 4.13   Failure modes for unidirectional composites subjected to compressive loading.

> *30° to 45°*: Combined matrix lacerations and cleavage
> *45° to 90°*: Extensive matrix cleavage
> *Compression*: The many modes of compression loading failure of composites are illustrated in Figure 4.13, and like tension loading, they are dependent on the

fiber direction. In longitudinal loading, the fibers support the load up to buckling. Microbuckling (short-wave stability) is representative of the failure modes of current graphite/epoxy composites, and forms of microbuckling are either in-phase buckling (kinking), in which the matrix shear stiffness is inadequate and the whole layer shears sideways, or out-of-phase buckling, in which the matrix direct transverse extensional stiffness is inadequate. The matrix supports transverse and normal compression loading, and so the failure process is dependent on the matrix stiffness. If the matrix is stiff enough to resist buckling, then fracture is by transverse shear cracking (see Figure 4.14).

Shear: The interlaminar shear and in-plane shear strengths of composite laminates are also matrix-controlled strengths. Matrix shearing action will initiate cracks in the lamina, and these will propagate parallel to the fiber direction. Interlaminar shear loading has a major bearing on the initiation of delaminations at free edges. Interlaminar crack propagation due to shear is the same as normal compression cracking (see Figure 4.14).

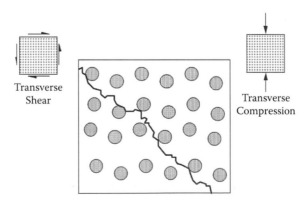

Figure 4.14   Matrix cracking under transverse shear or compression.

Failure Mechanisms

There are three basic stages of damage development in composite materials. These basic stages are crack initiation, crack growth, and localization of cracks leading to ultimate failure. The basic failure mechanisms of these stages of damage development are therefore:

1. Matrix cracking (delaminations and transverse)
2. Interface failure (fiber/matrix debonding)
3. Fiber fracture

These basic failure mechanisms correlate well with the in-service damage types discussed previously.

In an isotropic body, there are six possible crack orientations, as seen in Figure 4.15. With the introduction of fibers into such a body, certain crack orientations are preferred. Crack propagation tends to be along the lowest-strength path. Hence, planes of weakness in the composite lamina and composite laminate would develop. Because the matrix generally has a lower strength than the fibers in polymeric composites, crack orientation would be predominantly through the matrix, parallel to the fibers. Using Figure 4.15 as an illustration, if the fibers are parallel to the longitudinal

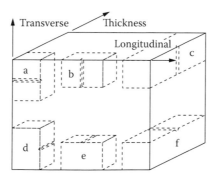

Figure 4.15 Crack orientation in an isotropic body.

direction, then cracks *a*, *c*, *d*, and *f* would be more likely to propagate.

Invariably, in multidirectional composites, the first mode of damage is matrix cracking, with the fibers acting as matrix crack arresters. Fortunately, the structure can develop microcracks in the matrix without endangering the structural integrity of the component. Impact damage is a good example of this, in that transverse matrix cracks and delamination form in a truncated pyramid pattern (Figure 4.16), but under tensile loading the strength is still quite high (typically >85% of the undamaged state).

The fracture behavior of composite laminates with through-the-thickness holes, and under in-plane loading, depends on:

Notch tip radius
Hole and panel geometry
Loading type
Laminate thickness
Ply orientation
Stacking sequence
Material properties

A phenomenon known as *hole-size effect* in composite laminates is where, under in-plane loading, a larger hole causes a greater stress reduction than a smaller hole, although the

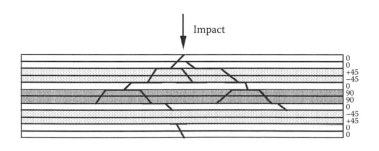

Figure 4.16   Matrix crack pyramid pattern from impact damage.

maximum in-plane stress on the boundary of the hole is the
same for all hole diameters. Moiré interferometry has indicated
that high out-of-plane deflections or buckling increase the edge
out-of-plane stresses. The fibers at the hole edge locally buckle,
and the damage propagates by shear crimping and delamination up to ultimate laminate failure. This buckling is more
severe when the ratio of hole diameter to panel width is small.
The effect is primarily due to interlaminar stresses induced
at the free edge, where cracks initiate and propagate in the
matrix. However, the final fracture mode does change from
one point to another around the hole boundary, and in the
end the failure tends to be less than that predicted by classical
lamination theory.

The fracture process of delaminations simply involves
interlaminar cracking between two highly anisotropic fiber-
reinforced plies. However, the fracture process is very
complex because it depends on material and geometric discontinuity, and it appears to involve coupling effects of the
three distinct modes of crack propagation: Modes I, II, and
III. Delaminations are initiated from either the free edge
out-of-plane stress induction or the growth of matrix cracks
to ply interfaces (Figure 4.17). The progressive fracture behavior of delaminations is shown in Figure 4.18, which illustrates
the characteristics of matrix tearing and hackle formation.

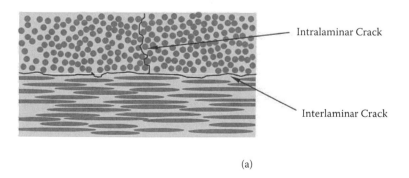

Intralaminar Crack

Interlaminar Crack

(a)

Figure 4.17a   Matrix cracks and delamination initiation.

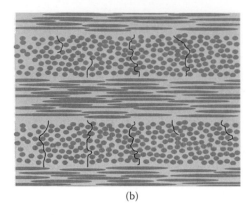

(b)

Figure 4.17b Matrix cracks and delamination initiation.

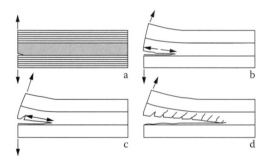

Figure 4.18 Delamination initiation and hackle formation.

Due to the low interlaminar strength of untoughened graphite/epoxy composites, delaminations are easily initiated through impact damage or simple out-of-plane–induced stresses under in-plane loading conditions.

The propagation of delaminations in graphite/epoxy composites is sensitive to:

Matrix toughness
Ply stacking sequence
Component shape and constraints
Delamination position

Delamination size
Load type and magnitude
Environmental effects

   Generally, the presence of a delamination reduces the overall stiffness of a composite structure. This lowers the critical buckling load and can result in local laminate structural instability under compressive loading. The final failure of a delaminated structure is by:

Increased net section stresses
Out-of-plane bending
Asymmetric twisting

Conclusions

Advanced polymeric composite materials are prone to a large number of defects and damage types. These defects and damage types emanate from the constituent material processes, composite component manufacture, and in-service use of the composite component. Those defect types due to in-service usage can be generalized further as:

Transverse matrix cracks
Delaminations
Holes (fiber fracture)

   The failure modes of composite materials are numerous and are influenced by many factors. In multidirectional laminates, the prediction of failure modes is very difficult and is usually a combination of several unidirectional failure modes under the various loading spectrums. In the majority of failure cases, the failure mode is determined by postmortem examination of the fracture surface.

Damage progression in composite components can be summarized as follows:

1. Multiple matrix cracking (termed the *characteristic damage state*) is the first stage of damage, where the damage pattern is random and scattered over the laminate.
2. The second stage of damage progression is the initiation of delaminations, or damage localization. Here the damage develops at preferred sites, such as free edges or ply interfaces.
3. The final fracture is often multimoded with severe cracking, but fiber fracture appears to be the controlling factor.

Composite fracture behavior for untoughened graphite/epoxy laminates in the presence of holes and delaminations is governed by the interlaminar integrity of the laminate.

Chapter 5

Loss of Integrity

Introduction

From the review of defect and damage types in composite and adhesively bonded joints (Chapter 2), we need to determine which defects are more serious in terms of structural integrity. The three generic defect or damage types that classify the majority of the 52 defect types in composite materials and 6 defect types in sandwich structures are matrix cracks (intralaminar), delaminations (interlaminar matrix cracks), and fiber fracture (holes). The two general defect types on adhesively bonded joints are debonds and weak bonds. The structural significance of these is discussed herein.

Definitions

This chapter presents a brief definition of the three types of generic defects found in composite structures and the two for adhesively bonded joints. For the sake of clarity, the chapter also discusses what is meant by the term *structurally significant defects*.

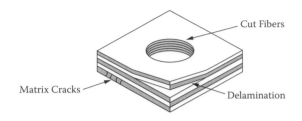

Figure 5.1   Principal defect types in composite structures.

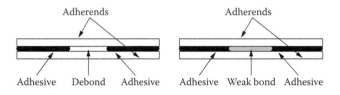

Figure 5.2   Typical defect types of adhesively bonded joints.

Generic Defect Types

The three generic defect types in composite structures are depicted in Figure 5.1, which clearly highlights their definitions. Figure 5.2 shows the two generic types of defects in adhesively bonded joints.

Structurally Significant Defects

Structurally significant defects or damage can be defined in terms of:

- Degradation of the structural strength and/or stiffness, and/or
- Defect instability under the in-service loading spectrum

General Representation

The general representation of all defects is listed in Tables 5.1 and 5.2 for those found in adhesively bonded joints and composite structures, respectively.

Table 5.1   Generalized Defect Types for Adhesively Bonded Joints

Debonds	*Weak Bonds*
Debond or unbond	Adhesion variability
Edge damage	Inclusion
Impact damage	Moisture pickup/aging
Inclusion	No fillet
Missing adhesive	Over/under cured
Porosity	Poor adhesion
Retained release film	Porosity
Voids	Retained release film
	Thermal stresses
	Variation in adhesive properties
	Variation in adhesive thickness

Criticality of Defects

The assessment of a defect's criticality is based on several factors. When assessing the criticality of a particular defect, each of these factors must be considered and their weighted importance judged accordingly. The influencing factors are as follows:

- The severity of the defect, taking into consideration its size, shape, volume, distribution, and nature
- The location and orientation of the defect as to whether it goes through all plies in the laminate or if it is related to one particular ply
- The frequency of the defect's occurrence
- Component load-path criticality and stress state
- How the defect is idealized in the assessment
- The local design load levels and their nature—for instance, tension, compression, shear, or a combination of the three—and whether the loading is in-plane or out-of-plane

Table 5.2 Generalized Defect Types for Composite Structures

Delaminations	Matrix Cracks	Holes	Design Variance
Bearing surface damage	Bearing surface damage	Bearing surface damage	Creep
Blistering	Contamination	Crushing	Damaged filaments
Contamination	Corner/edge crack	Cuts and scratches	Dents
Corner/edge crack	Cracks	Fastener holes	Erosion
Corner radius delamination	Edge damage	Fiber kinks	Excessive ply overlap
Delaminations	Matrix cracking	Fracture	Fiber distribution variance
Debond	Matrix crazing	Holes and penetration	Fiber faults
Edge damage	Porosity	Reworked areas	Fiber kinks
Fastener holes	Translaminar cracks	Surface damage	Fiber misalignment
Fiber/matrix debond	Voids		Marcelled fibers
Holes and penetration			Miscollination
Pills and fuzz balls			Mismatched parts
Surface swelling			Missing plies
			Moisture pickup

Nonuniform agglomeration of hardener agents	
Over-aged prepreg	
Over/under cured	
Pills and fuzz balls	
Ply underlap/gap	
Prepreg variability	
Surface oxidation	
Thermal stresses	
Variation in density	
Variation in resin fraction	
Variation in thickness	
Warping	
Wrong materials	

- Defect detection capabilities and detectability
- Local repair capabilities
- The component's configuration, which includes layup, thickness, joint type, and structural constraints
- Environmental conditions of the component
- The load history of the local area
- Variations in the material system, i.e., changes in material toughness due to moisture degradation

Table 5.3 shows the effect on the residual strength of a component where the three general defect types in composite structures are present. From this result, we can conclude that matrix cracks show no significant effect on strength degradation and that it is only delaminations and cut fibers that have a strength-reducing influence.

Matrix Cracks (Intralaminar)

As shown in Table 5.3, matrix cracks have no immediate effect on the component's strength. However, they should not be

Table 5.3   Effect of Local Damage on Residual Strength

	Resulting Local Damage		
	Cut Fibers	*Matrix Cracks*	*Delaminations*
Material or Manufacturing Error	Few	Few to very many	Minimal
Impact or Puncture	Many		Moderate
Battle Damage	Very large number		Extensive
Residual Strength	Above or below $p_{threshold}$, depending on number of cut fibers	No effect on residual strength	Above or below $p_{threshold}$, depending on delamination size

ignored either. Intralaminar matrix cracks do cause concern, and therefore they need to be repaired. The reasons for this concern are:

■ They open the component to further environmental degradation such as increased moisture degradation of composites and adhesives, corrosion of adherents and metal honeycomb core, and the induction of other harmful chemicals and fluids that are difficult to remove.
■ They are a source of delamination initiation, which further degrades the component strength and eventually can lead to ultimate or gross failure.
■ They are an area of local stiffness reduction and therefore can propagate by local structural instability.

Delaminations (Interlaminar Matrix Cracks)

The size and location of a delamination in composite materials and of a debond in adhesively bonded joints are the critical issues for strength reduction. Delaminations (which also cover debonds) can severely reduce the strength and stiffness of components. However, it is only under compressive loads, and to some lesser degree shearing loads, that this is degradation is true. When a delaminated composite material is subjected to tension, the residual strength is generally reduced by only 10% to 15%. The stiffness is only slightly affected due to asymmetric warping of the outer plies. Figure 5.3 illustrates the local warping of a delaminated composite panel under tensile loads.

The compressive loads in a delaminated composite component cause severe out-of-plane bending or buckling (see Figure 5.4), and these induce the out-of-plane interlaminar stresses that ultimately cause crack growth.

Figure 5.3 Holographic interferogram showing asymmetric warping of a delaminated composite plate under tension.

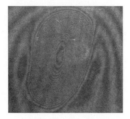

Figure 5.4 Holographic interferogram showing out-of-plane buckling of a delaminated region.

In adhesively bonded joints, the severity of the debond depends on the degree of damage tolerance that has been designed into the joints. Most bonded-joint overlaps provide a significant amount of damage tolerance, as shown in Figures 5.5–5.7.

Fiber Cuts and Holes

Whenever fibers are cut, which is generally in the presence of a hole, we have a concentration of stress distribution. The effect on the residual strength is significantly affected under all loading conditions, and the resulting stress concentration factor will indicate the severity of the hole.

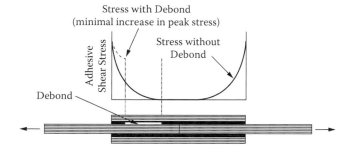

Figure 5.5   Internal edge debond in an adhesively bonded joint.

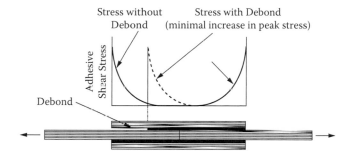

Figure 5.6   Edge debond in an adhesively bonded joint.

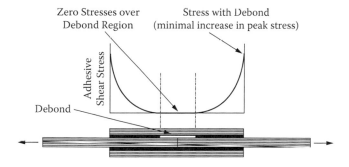

Figure 5.7   Central debond in an adhesively bonded joint.

Summary

From the previous discussions, it is clear that the defects posing the most concern to the structural integrity of a component during its service life are

- Intralaminar matrix cracks
- Delaminations (interlaminar matrix cracks) and debonds
- Holes

Chapter 6

Restitution and Repair

Introduction

Once the damage of a defect has been identified and evaluated against appropriate criteria to determine the loss of structural or performance integrity, a repair scheme is developed. This chapter examines the fundamentals of developing the repair scheme.

The development of a repair scheme will need to consider several practical implementation issues and review the specific repair requirements for successful design and installation. This chapter presents these repair design and installation requirements. The chapter also discusses the development of generic repair types in composite structures and describes repairs for the generalized defect/damage types (intralaminar matrix cracks, delaminations, and broken fibers).

The concluding section of the chapter provides an overview of the damage removal and repair scheme installation process. This final section also includes a detailed discussion on surface preparation requirements and processes.

Selection of the Repair Method

The repair design is primarily driven by nonengineering requirements such as:

■ The repair facility capability
■ The type of damage found
■ Whether or not the repair scheme is to be installed on or off the aircraft
■ The accessibility of the damaged area

Repair facility capability: The capability of suitable repair facilities has the strongest influence on the repair design. Without the appropriate tools, equipment, and materials at hand, even the best repair designs cannot be installed. The level of repair a facility is authorized to undertake is dependent on the capability of the appropriate facilities. For example, flight-line repairs are generally restricted to simple plug/patch repair types, whereas a depot-level repair should cover all repair types.

Types of damage: The types of damage that have already been reviewed in terms of their structural significance often dictate the repair design. The damage types discussed in terms of repair design will be

Matrix cracks
Delaminations
Debonds
Holes

On or off structure: When repairing a damaged structural component, the question arises of whether the component should be removed. Some components, such as access panels and doors, can be easily removed and replaced by a serviceable component. The damaged door can then be repaired in the repair facility, where appropriate environmental conditioning can take place.

However, external skins on wings, fuselages, etc., cannot be removed, and repairs must be done on the structure. There are two criteria that guide the decision to conduct repairs on the aircraft or to remove the component for repair:

1. A comparison of time to remove and install the component to the time to repair it
2. The hours to repair directly on the aircraft as opposed to removal, repair, and reinstallation of the component

Damaged component accessibility: The location of the damaged component is a serious limitation to repair design. If two-sided access is available, the repair design is often more effective. There are also specific application methods that allow repairs to damaged components when only one-sided access is available, and these are discussed later in this chapter.

Repair Criteria

The basis of the repair design follows a logical repair criterion. The parameters of the repair criteria are listed in Table 6.1 and are detailed in the following list.

Static strength and stability: Any repair must be capable of supporting the design loads that are applied to the original structure. The two major aspects of this are:

1. *Strength restoration*: The first question to ask is if full-strength restoration is required. The answer to this question is determined on the results of the damage analysis.
2. *Stability requirements*: The greatest concern in many of the damaged structures is instability under compressive loading and how to restore structural stiffness. The damage analysis will indicate where structural instability exists and specify the methods of designing a repair to overcome this instability.

Table 6.1 Repair Criteria

1. Static Strength and Stability
  • Full- versus partial-strength restoration
  • Stability requirements

2. Repair Durability
  • Fatigue loading
  • Corrosion
  • Environmental degradation

3. Stiffness Requirements
  • Deflection limitations
  • Flutter and other aeroelasticity effects
  • Load path variations

4. Aerodynamic Smoothness
  • Manufacturing techniques
  • Performance degradation

5. Weight and Balance
  • Size of the repair
  • Mass balance effect

6. Operational Temperature
  • Low- and high-temperature requirements
  • Temperature effects

7. Environmental Effects
  • Types of exposure
  • Effects to epoxy resins

8. Related On-Board Aircraft Systems
  • Fuel system sealing
  • Lightning protection
  • Mechanical system operation

Table 6.1 (*Continued*)   Repair Criteria

9. Costs and Scheduling
   • Downtime
   • Facilities, equipment, and materials
   • Personnel skill levels
   • Materials handling

10. Low Observables

Repair durability: Any repair designed to restore the aircraft to flying conditions is generally expected to remain an integral part of the airframe for the aircraft's remaining service life (exceptions are rapid-action-type repairs). For commercial aircraft, their serviceable life is 50,000 flight hours, and for military aircraft it is 4,000 to 6,000 flight hours, plus any life-of-type extensions. Thus, the durability of the repair scheme must consider the following in its design phase:

  – Fatigue loading of the structure and the effects on bolted and bonded joints, damage growth, and monitoring the repair for continuing airworthiness assessment
  – Corrosion of components where dissimilar materials have been used in the repair to ensure that corrosion protection precautions are still in place
  – Environmental degradation of resin-type repairs, particularly moisture absorption and performance in hot/wet environments

Stiffness requirements: In aircraft where lightweight structures are an essential design requirement, stiffness is often more critical than strength. The same goes for repairs, as they must maintain the integrity of structural stiffness. The following must be considered in a stiffness-repair design requirement:

  – Deflection limitations of flying surfaces such as wings and flight controls are based on the aerodynamic performance of the aircraft; repair should not unduly alter the aircraft's flying characteristics.

- Flutter and other aeroelasticity effects restrict the design of a repair so that its stiffness should be almost equal to the parent structure. Increased stiffness will decrease the flutter speed, and a decrease in stiffness can also change the flying characteristics.
- Load-path variations obviously are undesirable in that areas within the structure will be loaded in excess of their design allowables. As a general rule, the repair area stiffness should match that of the parent structure.

Aerodynamic smoothness: Aerodynamic smoothness is an important consideration when maximum speed or fuel efficiency is required. Those parts of the aircraft that require good aerodynamic smoothness, i.e., leading edges and where the boundary layer is laminar, must have flush or very thin external patch repair schemes. These repair types are based on local capabilities in manufacturing techniques, the effects of performance degradation, the repair size, and the possible effects of multiple damage sites.

Weight and balance: The size of the repair and the local changes in weight can be insignificant to the total component weight, but in weight-sensitive structures, such as flight control surfaces, the effect to the mass balance can be highly significant. The effective change in local weight must be controlled to within certain limits, and in some cases rebalancing of the component may be necessary.

Operational temperature: The operating temperature influences the selection of repair materials, particularly adhesives and composite resins. Materials that develop adequate strength within the required operational temperature range must be selected. The combination of extreme temperatures with environmental exposure, the hot/wet condition, is often the critical condition to which the repair must be designed.

Environmental effects: Composite and adhesive bonds are prone to significant degradation when exposed to various environments, in particular fluids and thermal cycling. However, absorbed moisture is frequently the major long-term concern in terms of durability of the repair design.

Related on-board aircraft systems: The repair design must also be compatible with other onboard aircraft systems. Typically these systems are as follows:

- Fuel system sealing: In modern aircraft, the fuel is carried within the wing as a "wet wing." Hence any repair to wing skins that are in direct contact with the fuel system must seal the fuel tank, cater for out-of-plane fuel pressure forces, and not contaminate the fuel system during the repair process.

- Lightning protection: If electrical conductivity in the parent structure has been required for lightning protection, then the repair must also incorporate the same degree of electrical conductivity.

- Mechanical system operation: Any component that is required to move during the operation of the aircraft or is in close proximity to a moving component, and is subsequently repaired, must ensure that the repair does not impede component operation. For example, retracting flaps must be repaired such that the repair still provides the adequate retraction clearance.

Costs and scheduling: Repairs and their design cost are considered in light of aircraft downtime and operating expenses. However, it is well established that it is cheaper to repair than replace, given appropriate facilities and adequate personnel skilled to do the repair.

Low observables: A most important attribute for today's military aircraft is its reduced radar cross-section (RCS), which is a major contributor to its stealth characteristics. If the aircraft is a stealth type of design, then repairs must be designed to maintain the mold line and not have reflective corners. For aircraft that have low stealth characteristics, like the F-16 and F-18 types, the repair of stealth design is not a concern.

Generic Repair Designs

There are four basic levels of generic repair designs. There are:

1. Filling and sealing the damaged area (cosmetic)
2. Filling and applying a doubler patch to the damaged area (semistructural)
3. Bonding a flush patch to the damaged area (structural)
4. Bolting a patch to the damaged area (structural)

Filling/Sealing Repair Scheme

When the significance of the damage is small and the main requirement to repair is for environmental protection, a cosmetic repair is all that is necessary. In such a repair scheme, the damage may not necessarily be removed, but moisture removal is still required. The damaged area is filled with a suitable potting compound (neat resin or mixed with chopped fiberglass) and then sealed with a layer of fiberglass/epoxy woven cloth (Figure 6.1).

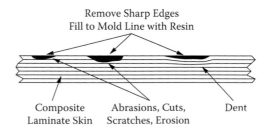

Figure 6.1 Cosmetic repair (nonstructural).

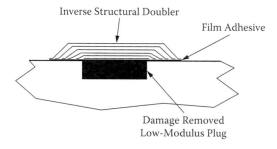

Figure 6.2 Semistructural plug-patch repair.

Filling/Doubler Patch Repair Scheme

As the structural severity of the damage increases, particularly for thin skins and honeycomb sandwich panels, some load transfer over the damaged region will be required. Such a repair scheme is both cosmetic and semi-structural (Figure 6.2). The damage is usually removed, and so is moisture; honeycomb core is replaced or a foaming resin plug inserted, over which a doubler is bonded. The plug must be of low modulus so as not to attract load; hence the load is transferred from the parent laminate into the doubler and out again into the parent laminate. The plug should not draw much load from the doubler.

Flush-Bonded Patch Repair Scheme

In relatively thin structures that have been significantly reduced in strength by the presence of damage, a flush-bonded patch repair scheme will provide the greatest strength restoration where aerodynamic smoothness is essential. The repair process is to remove the damage and carefully scarf or step the hole out. Again, drying the laminate prior to repair is important. The patch, designed and cut to fit in the hole, can by either pre-cured and secondarily bonded or co-cured to the damaged area. Co-cured patches are generally stronger. A doubler patch is also included in the repair scheme as a sealing patch for the flush patch. The doubler patch is no more than four plies in thickness and can be of a lower-modulus material. Figure 6.3 illustrates typical flush-bonded patch repairs.

Bolted-Patch Repair Schemes

The bolted-patch repair is restricted to thicker laminated sections that require ample structural integrity to be restored. Although the full structural strength is unlikely to be achieved, restoration of the design load-carrying capacity can be achieved with a bolted patch repair. The bolted patch repair can be a semi-flush or double-lap repair scheme, depending on the design requirements, and is typical of that shown in Figure 6.4. The repair process is to remove the damage and create a hole with circular ends, remove any moisture, drill the locating fastener holes in the parent laminate, and attach the inner, flush, and outer patch panels. The patch panels and fasteners should be coated with a sealing compound and fitted wet.

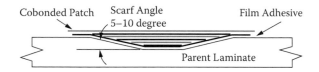

Figure 6.3 Flush-bonded patch repairs.

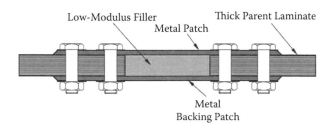

Figure 6.4 Bolted patch repair.

Matrix Cracks

We recall that matrix cracks have little effect on the structural strength integrity, but they can cause local stiffness losses and thus instability problems under compressive loading. Hence, there are two types of repairs required:

1. If the matrix cracks are insignificant as a damage type on the structural integrity of the composite laminate—and if they are exposed to the surface—then only a filling/sealing-type repair is warranted (Figure 6.1). This type of repair will ensure that moisture is excluded from the damaged area.
2. If damage analysis indicates that local structural instability is likely, then the damaged region is filled and sealed with a doubler patch to restore local stiffness (Figure 6.5). The $E_{restored}$ should be equivalent to the undamaged stiffness such that:

$$E_{restored_1} = \frac{E_{parent_1} b}{t_{repair}} = \frac{E_{patch_1} t_{patch} + E_{damage_1} t_{damage}}{t_{repair}}$$

$$= \text{Effective Patch Stiffness} + \text{Effective Degraded Stiffness}$$

where:

$E_{parent1}$ = parent laminate effective principle stiffness

b = laminate thickness

t_{repair} = thickness of repaired region

E_{patch1} = patch stiffness

t_{patch} = patch thickness

$E_{damage1}$ = damaged region stiffness

t_{damage} = depth of damaged region

$E_{restored1}$ = restored stiffness

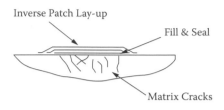

Figure 6.5   Doubler-patch installation.

In the repair scheme designs:

1. The damage is not cut out because load-carrying fibers are still in place.
2. Prior to repair scheme installation, the local damaged region is dried out.
3. A low-viscosity filling/sealing resin is used.

As a good design practice, doubler patches are stacked in reverse order with the largest ply on the top (Figure 6.5). This practice provides both a final seal to the repaired area and reduces peel stresses due to any load transferred through the patch.

Delaminations

Delaminations are more of a concern to the structural stability of composite laminates rather than strength degradation; hence only a fill/doubler patch repair is normally warranted. If the delamination effect on the structural integrity is within the damage-tolerant design strain allowables, then no repair is required; however, if the delamination is exposed to a free edge, then filling with a low-viscosity resin and sealing is necessary against environmental degradation. The structural repairs of internal and edge delaminations are as follows:

Internal delaminations: Those internal delaminations under compressive loading that are close to the surface—and

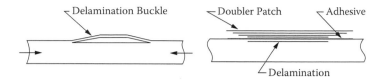

Figure 6.6 Doubler-patch installation over delamination.

where the damage analysis shows that the sub-laminate is likely to buckle under design-allowable strains—requires a doubler patch to increase the local stiffness of the sub-laminate (Figure 6.6). Determination of the patch stiffness is based on the previous analysis, but here the stiffness of the sub-laminate and patch needs to be such that the critical buckling load is greater than the applied design allowable load. Or, more simply, the stiffness of the patch must be equivalent to the undamaged stiffness, such that

$$E_{\text{patch}} = \frac{F_{\text{parent}}h}{t_{\text{patch}}}$$

Edge delamination: With an edge delamination, the first requirement of the repair is to seal the edge from further moisture absorption, and again a low-viscosity resin is used. Local stiffening of the edge is more difficult, since the driving out-of-plane forces are still present. The most effective repair design is to simply reinforce the out-of-plane direction. Since the out-of-plane stresses are much lower than in-plane, a fastener or thin capping patch is all that is required (Figure 6.7).

If the delaminations are severe, then the damaged region will have to be removed, and the repair scheme will then be for that of a hole. Prior to repair scheme installation, the damaged region will require the moisture to be removed.

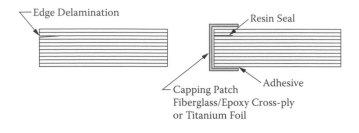

Figure 6.7 Out-of-plane reinforcing capping patch over edge delamination.

Holes and Fiber Fracture

Low-Strength Degrading Holes

Where the hole in the laminate presents a low-strength degradation to the overall laminate structural integrity, the general repair is a plug/patch scheme, as shown in Figure 6.2. For a rapid type repair, the damage is not necessarily removed, but in most cases it will be removed. The amount of damage removed should be minimal and be of simple geometric design, i.e., circle or circular ended. The plug and patch should be of a lower modulus than the parent material so that the repair area does not attract load. With a low-modulus plug, the stress concentration factor will reduce due to hole deformation constraints, and the patch is mainly used as a sealing cover over the plugged hole.

Moderate-Strength Degrading Holes

In the case where the damage analysis of a hole in a composite laminate indicates that there is moderate strength degradation, i.e., where the current level of damage tolerance is significantly reduced but catastrophic failure would only occur with severe overload, then a plug and structural doubler patch is recommended (Figure 6.8). Again, the plug is of low modulus so that the load path is from the parent laminate into

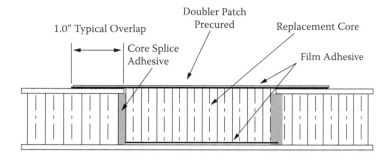

Figure 6.8   Structural doubler and plug repair.

the doubler, but not into the plug. The design of the patch follows a simple method, such that the conditions are as follows:

1. Half a double-lap joint is designed, acknowledging supports for bending resistance.
2. A tapered patch is used to reduce peel, particularly when the thickness of the patch is greater than 1 mm (eight plies).
3. The patch stiffness and thermal expansion coefficients are matched.
4. The hole is not tapered.

Fully Structural Repairs to Holes

When the hole causes a significant reduction to the laminate strength, a fully structural restoring repair is required. The repair will either be a scarf (stepped-lap) bonded patch for thinner structures; for thicker sections, a bolted patch is required.

> *Thin-section flush repairs*: A flush repair to a thin laminated section will either be a scarf joint or stepped-lap. Generally, the damage removed results in a stepped hole; however, using a scarf joint analysis, the final repair design will suit both joint types. In the simple analysis, we try to maintain stiffness and thermal coefficient of

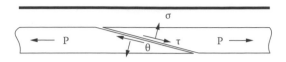

Figure 6.9   Scarf joint analysis geometry.

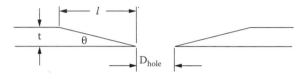

Figure 6.10   Scarf repair length geometry.

expansion balance. Therefore, if the load is acting over a scarf angle of $\theta°$, the normal and shear stresses are as shown in Figure 6.9. Using Figure 6.10 and a function of the laminate thickness for a hole size of 2 inches, Table 6.2 illustrates the increasing patch size if a scarf joint is used, clearly showing that this repair design is limited to thinner sections.

Thick section bolted repairs: As the skin thickness increases, scarf repairs become impractical in terms of repair design. At this stage, a bolted repair is more practical. The following design points are recommended for a bolted repair design:

1. Attempt to use a low-modulus plug that restricts the load into the filled hole, as a loaded hole has a greater strength reduction than an open hole.
2. Where a patch is relatively thick, taper the edges or use a stepped-lap configuration, as this will reduce the load on the first row of fasteners, which are the critically loaded ones.
3. Seal the patch to the parent laminate and install the fasteners with sealant in the wet condition.

Table 6.2 Scarf Patch Size to Laminate Thickness

No. of Plies	Laminate Thickness (in.)	Scarf Length (in.)	Patch Length (in.)	Lpatch/Dhole
8	0.04	0.76	3.52	1.8
12	0.06	1.15	4.30	2.2
16	0.08	1.53	5.06	2.5
24	0.12	2.29	6.58	3.3
36	0.18	3.43	8.86	4.4
50	0.125	4.77	11.54	5.8

Damage Removal and Surface Preparation

For successful installation of the repair scheme, the damaged area and potentially degrading locally induced environment need to be effectively prepared.

Removing the Damage

The basic principles of damage removal are to first make the installation of the repair scheme relatively easy to accomplish but also to avoid inducing poor load paths and stress-intensity areas of load from the parent structure through the patch and back out into the parent structure. Hence simple geometric patterns for the damage removed are highly recommended. Such simple geometric patterns include circular holes and domed-ended rectangles (Figure 6.11). Tight-radius cutouts should be avoided, as the tight radial produced, for example, with an elliptical cutout can induce high stress concentrations.

The repair scheme should always attempt to minimize the amount of undamaged structure removed. Thus straight-edged holes that would incorporate a doubler repair are ideal for this requirement. However, for flush repairs, the scarfing of the parent structure away from the damaged hole site requires specific angles of the scarf. As a rule of thumb, a 1° scarf

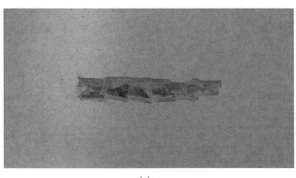

(a)

Figure 6.11a   Damaged composite sandwich panel skin.

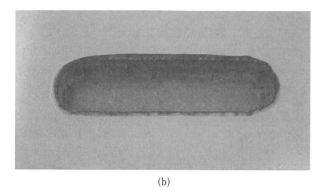

(b)

Figure 6.11b   Domed-ended rectangular cutout.

angle that corresponds to a ¼-in. per ply for a 0.005-in. ply thickness is common practice (see Figure 6.12). A ⅛-inch per ply scarf slope giving a 2.5° scarf angle is also acceptable. Poor damage removal can result in extensive rework of the damage/repair site and can significantly increase the repair scheme size (Figure 6.13).

Moisture Removal

Moisture ingress will have a significant effect on the installation of the repair scheme in two ways. Firstly, the uncured resin of the composite patch and the adhesive is hygroscopic, and any absorbed moisture in the resin will reduce the resin and adhesive properties. This problem can be eliminated by ensuring that the uncured resins and adhesives are not open to the atmosphere for extensive lengths of time (typically less than one hour, but the specific time is dependent on the resin system). Materials that have been in cold storage must be brought up to temperature while still wrapped in the protective packaging. The package is only opened when the material is up to room temperature.

The second issue with moisture at the repair site is the form of the moisture in the parent composite laminate or sandwich structure. Sandwich structure with trapped water in the core

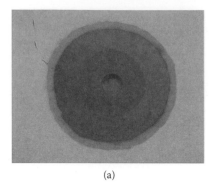

(a)

Figure 6.12a   Scarf slope of ¼ in. per ply.

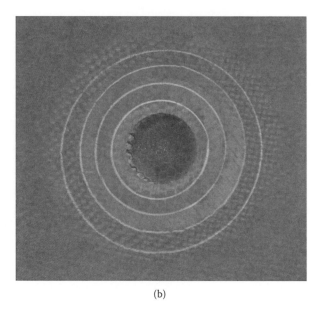

(b)

Figure 6.12b   Scarf slope of ⅛ in. per ply.

(especially honeycomb core) can create serious repair installation problems. Particularly with heat-cured repairs, the standing water in the core cells will boil and expand (Figure 6.14). There are several instances where the skin of the parent material has blown off the core as a result of this water expansion. Moisture absorbed in the parent composite laminate will also

Figure 6.13   Poor damage removal during scarfing.

Figure 6.14   Standing water in honeycomb cells of sandwich panels.

evaporate off during the heat-cured installation of the repair scheme. This evaporating water will create porosity in the repair scheme bondline, as shown in Figure 6.15.

Prior to repairing a composite laminate when using heat-cured adhesives, the parent laminate must be dried. Drying is normally done at 150°F–180°F, but can be higher if vacuum is used. The drying time is dependent on the drying temperature and the relative thickness of the laminate. Figures 6.16 and 6.17 illustrate the moisture absorption/desorption time as a function of temperature and laminate thickness, respectively. Best practice is to place the component to be repaired into an oven while developing the repair procedure or,

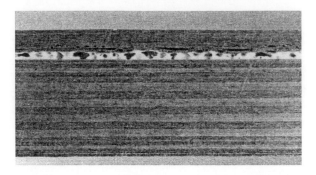

Figure 6.15 Bondline porosity due to moisture-contaminated parent composite laminate (from *Advanced Composite Repair Guide*, 1982).

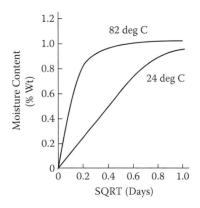

Figure 6.16 Temperature effects on the rate of moisture absorption.

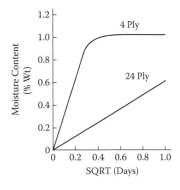

Figure 6.17 Laminate thickness effects on the rate of moisture absorption.

for components that can be removed from the larger structure, using a vacuum bag and heater blanket with temperature control.

Surface Preparation

Surface preparation of the adherend is critical for the durability or longevity of the bonded joint. The surface preparation of the adherend will depend significantly on the adherend type. In the case of composite materials, the surface preparation requirements are less stringent than those of metals.

Adherend Preparation (Surface Conditioning)

The key to effective adhesive bonding is a well-prepared adherend surface. The bondline is a complex interaction between the adhesive and adherend. It is made up of several interacting layers, as illustrated in Figure 6.18.

The bonding mechanism itself is quite complex, but it is essentially a combination of secondary bonding forces (van der Waals forces, Figure 6.19) and mechanical interlocking (Figure 6.20). Both of these primary bonding mechanisms are achieved through surface preparation. The electrostatic bonding is a result of surface cleanliness and adherend surface electron removal. Mechanical interlocking is achieved through surface roughening. Beyond these two types of surface preparation for bonding mechanisms, the addition of a surface-coupling agent can also be used. The surface-coupling

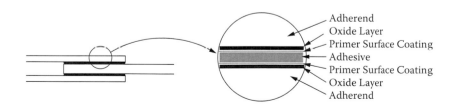

Figure 6.18 Adhesive bondline structure.

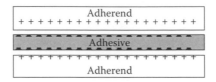

Figure 6.19   Effect of van der Waals force (electrostatic bonding).

Figure 6.20   Effect of mechanical interlocking.

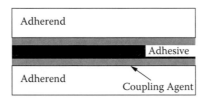

Figure 6.21   Effect of chemical bonding.

mechanism provides a chemical bond between the adherend and the adhesive independently (Figure 6.21). Surface-coupling mechanisms are more applicable to metal adherends than composite adherends.

The surface preparation process depends on the type of adherend material being bonded. For the common aerospace materials, the process is as follows:

   1. Composite adherend (Figures 6.22–6.24):
       a.   Removal of the peel ply, as voids are likely to result
       b.   Solvent cleaning to remove oil and grease
       c.   Grit blasting (aluminum oxide preferred) to roughen the surface

Figure 6.22   Peel-ply removed surface at 30× magnification (courtesy L.J. Hart-Smith).

Figure 6.23   Peel-ply removed and hand-sanded surface at 30× magnification (courtesy L.J. Hart-Smith).

Figure 6.24   Peel-ply removed and grit-blasted surface at 30× magnification (courtesy L.J. Hart-Smith).

Figure 6.25   Poor surface-preparation outcome.

2. Titanium patch:
   a. Solvent cleaning and abrasion
   b. Chemical etching (PASA jell) or silane pretreatment
   c. Corrosion-inhibiting primer (uniform layer)
   d. Preferably done in workshop, not field
3. Aluminum structure:
   a. Degreasing with solvent
   b. Grit blasting or mechanical abrasion
   c. Etching, if possible, or silane pretreatment
   d. Primer

Poor surface preparation of metal structures will typically be seen as one side of the fracture bondline showing the entire adhesive and the other side of the bondline as a clean metal surface (Figure 6.25).

Repair Scheme Fabrication and Application

Following the design of the repair scheme and removal of the damaged area, the repair patch (scheme) is fabricated and installed on the parent structure. The size of the repair scheme is very much dependent on the damage removed and may require a quick review of the repair scheme size due to an increase in the size of the damage removed. As part of the design of the repair scheme, the adhesive would have been selected. If the adhesive is a cold-storage film adhesive that will be cured at high temperature, it must removed from the freezer and allowed to come up to room temperature before opening the protective package. If the adhesive is a two-part mixed adhesive, the preparation of the curing-process materials must be done before the adhesive is mixed.

The following aspects must be considered in the development of the repair installation:

1. Simplicity of the repair:
   a. A simple repair is easier to install and thus reduces the potential of installation errors.
   b. Achieving the design requirements is more likely if a simple repair reduces the likelihood of installation error.
   c. Alignment of the honeycomb core ribbon direction is a good installation practice. In a few circumstances, the alignment direction of the core ribbon will maintain the parent structural and operational performance.
   d. A low-modulus plug is best included in the repair if load transfer into the plug is not desired or designed for. The low-modulus plug is installed to avoid attracting a load.
   e. The precision of the repair patch alignment to the parent structure is very much dependent on the load level being transferred through the repair patch and the magnitude of the parent/repair stiffness properties. Misalignment of the repair patch will require engineering disposition to determine if the repair scheme is to be accepted or reworked.
   f. The configuration of the repair's ply-stacking sequence is best mirrored by the parent structure for a doubler repair to maintain stack balance. For scarf repairs,

the repair ply configuration is best laid up to match the parent stack, as this arrangement will provide the more effective property restoration. See Figure 6.26 for the repair doubler arrangement and Figure 6.27 for a scarf repair arrangement.

2. A cover ply is often included on the repair patch scheme to act as an environmental seal. This ply is best made of a low-modulus material and configuration so as not to influence the engineering properties of the repair patch.

Cure Process

There are three important cure-process functions to consider when repairing defects/damage in composite structures. These three cure-process functions are the placement of the

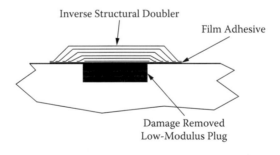

Figure 6.26   Repair ply-stacking sequence: doubler repair patch stacking sequence.

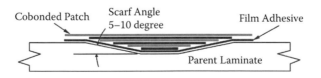

Figure 6.27   Repair ply-stacking sequence: scarf repair patch stacking sequence.

thermal-couple wires, a check for vacuum bag leaks, and a further check for a difference between positive and vacuum pressure during the process. Each of these three process functions are discussed as follows:

1. *Thermal-couple placement*: Thermal couples are placed in the repair assembly to both monitor and regulate (control) the cure temperature. Ideally the temperature over the repair assembly would be uniform, but this is not the case in reality. Variation in the heater blanket's thermal pattern, as well as repair structure heat sinks, will set up a varying thermal profile. Hence, prior to undertaking the actual repair, a thermal survey of the repair area is highly recommended. The thermal survey will indicate the area that is the coldest (due to a heat sink) and the hottest area (typically a thinner section of the structure). The hottest area is used to control the maximum temperature, and the cold spot can be used to determine the length of the cure time.

2. *Bag-leak check*: Prior to running the cure cycle of the repair, it is important to run a leak check of the vacuum bag. If an existing leak is not identified in the vacuum bag, there is a significant chance of the repair patch and adhesive bondline showing major porosity. The vacuum leak check is such that if a vacuum dial gauge loses pressure at 1 in. Hg over 60 seconds, then the leak check has failed.

3. *Positive pressure versus vacuum pressure*: Surface pressure over the repair area can be maintained by the vacuum itself (1 atmosphere). Note that the magnitude of the vacuum pressure is a function of the local air pressure. Positive pressure can also be applied through mechanical or pressure-vessel (autoclave) systems. Positive pressure is typically better for control of porosity and thickness (fiber-volume ratio) than vacuum pressure.

Post-Repair Inspection

Following restitution of the defect/damage, an inspection of the repaired area is highly recommended. Inspection of the repaired area provides evidence of appropriate restoration action. This inspection process will depend on the NDI equipment available and the skills of the NDI technicians. However, at a minimum, the repair area needs to be inspected visually and tap tested.

Visual inspection of the repaired area entails looking for obvious signs of resin flow at the edge of the repair scheme (Figure 6.28). Bleed evidence during repair teardown is also a visual inspection process. The visual inspection should also identify the quality of the fillet and establish whether the resin is still tacky. Evidence of resin flow in the repair patch (in the bleeder cloth) is part of the visual inspection process (Figure 6.29).

The second NDI post-repair method is simply the tap hammer test. The repair region is lightly tapped both around the region periphery and through the center of the repair patch. The tap test is an attempt to ensure bondline integrity by listening for the quality of the sound level. If tapping produces a relatively dull sound when compared to other areas of the repair patch, then this is good of indication of a potential debond.

There are several more detailed NDI methods that can be used to evaluate the integrity of the repair bondline.

Figure 6.28   Repair post-cure resin flow.

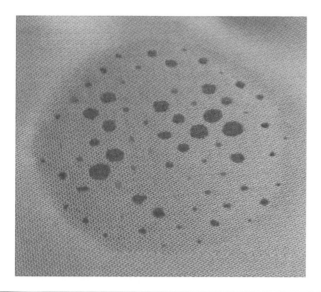

Figure 6.29   Repair patch bleed evidence in bleeder.

NDI techniques such as ultrasonics, optical methods, and thermography can also be used. These methods are discussed in detail in Chapter 3.

Summary

The restitution of a defect or damage requires a good appreciation of the severity to determine the best repair scheme for the composite structure. Several factors impact the repair scheme selection and can drive the best structurally efficient repair design to a more complex repair design, e.g., using a flush repair because of aerodynamics when a doubler repair would be ideal for strength restoration.

The defect/damage restoration can range from a simple low-viscosity resin infusion to a complex bolted repair scheme. Defect/damage restoration is an important requirement in the initial design of the structure. If the structure cannot be made damage tolerant, then it needs to be inspectable. If defects/damages are found, then it needs to be designed for repair.

References and Bibliography

References

ASM International. 1988. Composites. In *Engineered materials handbook*, Vol. 1. Metals Park, OH: ASM Int.

ASM International. 1990. Adhesives and sealants. In *Engineered materials handbook*, Vol. 3. Metals Park, OH: ASM Int.

Chen, H., and W. Chang. 1994. Delamination buckling analysis for unsymmetric composite laminates. In *Proceedings of the 39th International SAMPE Symposium*, 2855–67. Anaheim, CA: SAMPE.

Heslehurst, R.B. 1991. Evaluation of damage analysis techniques for composite aircraft structures. Master of Engineering thesis, Royal Melbourne Institute of Technology, Melbourne.

Heslehurst, R.B., and M.L. Scott. 1990. Review of defects and damage pertaining to aircraft composite structures. *J. Composite Polymers* 3 (2): 103–133.

Hoskin, B.C., and A.A. Baker, eds. 1986. *Composite materials for aircraft structures*. AIAA Education Series. New York: American Institute of Aeronautics and Astronautics.

Hsu, D.K. 2008. Nondestructive inspection of composite structures: Methods and practice. Paper presented at the 17th World Conference on Non-Destructive Testing, Shanghai, China.

Lagace, P.A., and N.V. Bhat. 1993. On the prediction of delamination initiation. In *Proceedings of the international conference on advanced composite materials: Advanced composites '93*, 335–41. Warrendale, PA: Mineral, Metals & Materials Society.

Lessard, L.B., and B. Liu. 1992. Fatigue damage simulation of a laminated composite plate with a central hole. In *Proceedings of the 24th International SAMPE Technical Conference*, 473–83. Toronto, ON: SAMPE.

Summerscales J., ed. 1987. *Non-destructive testing of fibre-reinforced plastics composites*. Vol. 1. London: Elsevier Applied Science.

Ullman, D. G. 2009. *The mechanical design process*. 4th ed. New York: McGraw-Hill.

Wang, A.S.D., M. Slominana, and R.B. Bucinell. 1985. Delamination crack growth in composite laminates. In *Delamination and debonding of materials*, ASTM STP 876, 135–67. West Conshohocken, PA: ASTM.

Williams, J.F., D.C. Stouffer, S. Ilic, and R. Jones. 1986. An analysis of delamination behaviour. *Composite Structures* 5:203–16.

Bibliography

"A Micromechanical Fracture Mechanics Analysis of a Fiber Composite Laminate Containing a Defect." V. Papaspyropoulos, J. Ahmad, and M.F. Kollninell. In *Effect of Defects in Composite Materials*. ASTM 836, 1984.

"A Quasi-Static Treatment of Delamination Crack Propagation in Laminates Subjected to Low Velocity Impact." T.C. Sun and C.J. Jih. Paper presented at 7th ASC Conference, October 1992.

"A Study on the Thermal and Moisture Influences on the Free-Edge Delamination of Laminated Composites." M.A. Mahler. Georgia Institute of Technology, September 1987.

"A Workshop on Practical Adhesive Bonding for Performance and Durability." M. Davis. ICCM-11, July 1997.

"Addendum to Design Manual for Impact Damage Tolerant Aircraft Structure." M.J. Jacobson. AGARD-AG-238 (Addendum) Report, 1988.

"Adhesive-Bonded Joints for Composites: Phenomenological Considerations." L.J. Hart-Smith. Douglas Paper 6707, presented at the Conference on Advanced Composites Technology, Technology Conferences Associates, California, March 1978.

"Adhesives and Sealants." In *Engineered Materials Handbook*, Vol. 3. Metals Park, OH: ASM Int., 1990.

Advanced Composite In-Service Damage Assessment/NDI Development Program. D.W. Nesterok. NAEC MISC 92-0366, September 1978.

Advanced Composite Repair Guide. NOR 82-60, Northrop Corporation, Hawthorne, CA, March 1982.

Advanced Composite Structures: Fabrication and Damage Repair. Abaris Training Course Notes, 1998.

"Advanced Damage Assessment, Stress Analysis and Repair of Composite Structures." R. Heslehurst. Abaris Training Course Notes, 2010.

"Advanced Instrumentation and Measurements for Early Nondestructive Evaluation of Damage and Defects in Aerostructures and Aging Aircraft." J.D. Achenbach, I.M. Daniel, and S. Krishnaswamy. AFRL-SR-BL-TR-98-0702, 1998.

Advanced Residual Strength Degradation Rate Modelling for Advanced Composite Structures. D.E. Pettit, K.N. Lauraitis, and J.M. Cox. AFWAL-TR-79-3095, August 1979.

"Advances in Evaluation of Composites and Composite Repairs." J. Tyson. Paper presented at 40th SAMPE Symposium, 1995.

"An Account of One Engineer's Long-Term Involvement with Aerospace Applications of Composite Structures, Part I: Analytical Developments." L.J. Hart-Smith. Boeing Paper PWDM05-0089, presented at the 37th ISTC SAMPE Meeting, Seattle, Washington, 2005.

"An Analysis of Delamination Behaviour." J.F. Williams, D.C. Stouffer, S. Ilic, and R. Jones. *Composite Structures* 5:203–16, 1986.

"An Analysis of Free-Edge Delamination in Laminated Composites under Uniform Axial Strain." C.S. Hong and K.S. Kim. In *Progress in Science and Engineering of Composites*, 261–68. Proc. of the 4th Int. Conf. on Composite Materials, Tokyo, 1982.

"An Experimental and Analytical Treatment of Matrix Cracking in Cross-Ply Laminates." S.E. Groves, C.E. Harris, A.L. Highsmith, D.H. Allen, and R.G. Novell. *Experimental Mechanics* 27:73–79, 1987.

Analysis and Performance of Fiber Composites. B.D. Agarwal and L.J. Broutman. New York: Wiley-Interscience, 1980.

"Analysis of Composite Laminates with Matrix Cracks." S.W. Lee and J. Aboudi. CCMS-88-03, 1988.

"Analysis of Compression Failures in Fibre Composite Laminates." R. Jones and R.J. Callinan. In *Progress in Science and Engineering of Composites*, 447–54. ICCM-IV, Tokyo, 1982.

"Analysis of Delamination in Compressively Loaded Laminates." D. Shaw and M.Y. Tsai. *Composites Science and Technology* 34:1–17, 1989.

"Analysis of F-18 Aircraft Composite Repair Requirements as Related to Ground Support Equipment." D.W. Nesterok. NADC-92-130, 1977–1979.

"Analysis of Local Delaminations and Their Influence on Composite Laminate Behaviour." T.K. O'Brien. In *Delamination and Debonding of Materials*, 282–97. ASTM STP 876, November 1983.

"Analysis of Plates." C.W. Bert. In *Composite Materials*, Vols. 7 and 8, *Structural Design and Analysis*, Chap. 4. New York: Academic Press, 1975.

"Analysis of the Effect of Matrix Degradation on Fatigue Behavior of a Graphite-Epoxy Laminate." R.T. Arenburg. Thesis, Texas A&M University, 1982.

"Analytical Modeling and ND Monitoring of Interlaminar Defects in Fiber-Reinforced Composites." R.L. Ramkumar, S.V. KulkArni, R.B. Pipes, and S. Chatterjee. In *Fracture Mechanics*, 668–84. ASTM STP 677, American Society for Testing and Materials, 1979.

"Application and Interpretation of a Holographic Interferometry Technique in the Detection of Defects in Structural Materials." R.B. Heslehurst. Thesis, UNSW, 1999.

"Assessment of Impact Damage in Composite Structures." R. Jones, S.C. Galea, and J.J. Paul. ARL-TR-23, 1993.

"Basic Failure Mechanisms in Advanced Composites." J.V. Mullin and V.F. Mazzio. NASA-2929-N03, April 1971.

"Behaviour of Advanced and Composite Structures." L.W. Rehfield. AFOSR-TR-85-0137, March 1983.

"Buckling of a Sublaminate in a Quasi-Isotropic Composite Laminate." K.N. Shivakumar and J.D. Whitecomb. NASA-TM-85755, February 1984.

"Characteristics of Free Edge Delamination in Angle-Ply Laminate." K.S. Kim and C.S. Hong. Conf. Proc. of the 5th Int. Conf. on Composite Materials, San Diego, CA., 347–61, 1985.

"Characterization, Analysis and Significance of Defects in Composite Materials." S.M Bishop and G. Dorey. Paper presented at 56th Meeting of the Structures and Materials Panel in London, AGARD-CP-355, April 1983.

Characterization and Analysis of Defects in Nonlinear Optical Molecular Composite Films. R. Mehta and C.Y.-C. Lee. WL-TR-92-4004, Report, 1992.

Characterization and Measurement of Defects in the Vicinity of Fastener Holes by Nondestructive Inspection. P.F. Packman, R.M. Stockton, and J.M. Larsen. AFOSR-TR-76-0400, September 1975.

"Characterization and Modelling of Impact Damage Growth in Adhesively Bonded Joints." R.S. Choudhry. Thesis, Department of Mechanical Aerospace and Civil Engineering, The University of Manchester, 2006.

Characterization of Defects in Composites and Their Effect on Mechanical Strength. G. Angell and M.F. Markham. NPL Rpt. A(A)20, December 1980.

Characterization of Deformation and Damage in Brittle-Matrix Composite Materials. I.M. Daniel. AFOSR-TR-96, PDF, May 1996.

Characterization of Delamination and Transverse Cracking in Graphite/Epoxy Laminates by Acoustic Emission. A. Garg and O. Isha. NASA-TM-84370, May 1983.

"Characterization of Impact Damage Development in Graphite/ Epoxy Laminates." In *Fractography of Modern Engineering Materials: Composites and Metals*, edited by J.E. Masters and J.J. Au, 238–58. ASTM STP 948, 1987.

Characterization of Manufacturing Defects Common to Composite Wind Turbine Blades: Flaw Characterization. T.W. Riddle, D.S. Cairns, and J.W. Nelson. *Papers: American Institute of Aeronautics and Astronautics* 2:837–54, 2011. http:// www.coe.montana.edu/composites/documents/2011_ AIAA_2011_1758%20Riddle.pdf.

"Characterization of Mode I and Mixed-Mode Delamination Growth in T300/5208 Graphite/Epoxy, Delamination and Debonding of Materials." R.L. Ramkumar and J.D. Whitecomb. In *Delamination and Debonding of Materials*, edited by W.S. Johnson, 315–35. ASTM STP 876, 1985.

"Characterization of Pure and Mixed Mode Fracture in Composite Laminates." T.E. Tay, J.F. Williams, and R. Jones. Dept. of Mechanical and Industrial Engineering, University of Melbourne, Australia.

Characterizations of Manufacturing Flaws in Graphite/Epoxy.
    G.A. Hoffman and D.Y. Konishi. Army Materials and Mechanics
    Research Center Report AMMRC MS 77-5, 1977.

"Classification of Composite Defects Using the Signature
    Classification Development System." J.S. Lin, L.M. Brown, and
    C.A. Lebowitz. In *Technology Showcase: Integrated Monitoring,
    Diagnosis and Failure Prevention.* Proceedings of a Joint
    Conference, Mobile, Alabama, 1996.

"Commercial Transport Aircraft Composite Structures: Failure
    Analysis and Mechanisms of Failure of Fibrous Composite
    Structures." J.E. McCarty. In *Proc. of a Workshop on Failure
    Analysis and Mechanisms of Failure of Fibrous Composite
    Structures*, 7–66. NASA-CP-2278, March 1982.

"Common Failure Modes for Composite Aircraft Structures due to
    Secondary Loads." A.M. Rubin. *Composites Engineering* 2 (5–7):
    313–20, 1992.

"Comparative Evaluation of Impact Performance of Laminate
    Composites Containing Impact Induced/Mechanically Inserted
    Defects." K. Kishore and B. Khan. *Journal of Reinforced Plastics
    and Composites* 17 (16): 1463–71, 1998.

"Composite Bolted Joints Behaviour: Effects of Hole Machining
    Defects." G. Gohorianu, R. Pique, F. Lachaud, and J.-J. Barrau.
    In *Joining Plastics 2006*, London, April 2006.

Composite Damage Tolerance and Maintenance Safety Issues, edited
    by L. Ilcewicz. FAA Workshop, 2006.

Composite Defect Criticality. S.N. Chatterjee and R.B. Pipes. NADC-
    82038-60, 1983.

"Composite Laminates with Elliptical Pin-Loaded Holes." E. Persson
    and E. Madenci. *Engineering Fracture Mechanics* 61:279–95,
    1998.

"Composite Materials: Processing, Quality Assurance and Repair."
    C. Zweben and M.G. Bader. Short Course Lecture Notes, UCLA,
    September 1983.

Composite Materials: Testing and Design (7th Conf.), edited by
    J.M. Whitney. ASTM STP 893, April 1984.

Composite Materials for Aircraft Structures, edited by B.C. Hoskin
    and A.A. Baker. New York: AIAA Education Series, 1986.

Composite Plates Impact Damage: An Atlas, edited by S.R. Finn and
    G.S. Springer. Lancaster, PA: Technomic, 1991.

Composite Structures Fabrication and Damage Repair: Phase 1.
    Abaris Training Course Notes, 2009.

"Composites." In *Engineered Materials Handbook*, Vol. 1. Metals Park, OH: ASM Int., 1988.

"Composites Damage Tolerance." W. Gregg. MSFC Conference Paper, 2008.

"Compression, Flexure and Shear Properties of a Sandwich Composite Containing Defects." A.P. Mouritz and R.S. Thomson. *Composite Structures* 44:263–78, 1999.

"Compression Behaviour of +\−45°-Dominated Laminates with a Circular Hole or Impact Damage." M.J. Shuart and J.G. Williams. *AIAA Journal* 24 (1): 115–22, 1986.

"Compression Behaviour of Unidirectional Fibrous Composites: Compression Testing of Homogeneous Materials and Composites." J.H. Sinclair and C.C. Chamis. In *Compression Testing of Homogeneous Materials and Composites*, edited by R. Chait and R. Papirno, 155–74. ASTM STP 808, 1983.

Compression Failure Mechanisms in Unidirectional Composites. H.T. Hahn and J.G. Williams. NASA-TM-85834, August 1984.

"Compression Failure of Composite Laminates: Failure Analysis and Mechanisms of Failure of Fibrous Composite Structures." R.B. Pipes. In *Proc. of a Workshop on Failure Analysis and Mechanisms of Failure of Fibrous Composite Structures*, 265–92. NASA-CP-2278, March 1982.

"Compression of Laminated Composite Beams with Initial Damage." N.J. Breivik, Z. Gurdal, and O.H. Griffin. Paper presented at 7th ASC Conference, 1992.

"Compressive Strength in Fiber-Reinforced Composite Materials." J.G. Davis, Jr. In *Composite Reliability*, 364–77. ASTM STP 580, 1975.

Compressive Strength of Carbon Fibre Reinforced Plastics, The. K.F. Port. RAE-TR-82083, August 1982.

"Compressive Strength of Composite Laminates with Interlaminar Defects." J.W. Gillespie Jr. and R.B. Pipes. *Composite Structures* 2:49–69, 1984.

"Compressive Strength of Damaged and Repaired Composite Plates." S.R. Finn and G.S. Springer. Paper presented at AIAA Conference, 1991.

"Consideration of Failure Modes in the Design of Composite Structures." L.B. Greszczuk. Paper no. 12 in *Failure Modes of Composite Materials with Organic Matrices and Their Consequences on Design*. AGARD-CP-163, October 1974.

Contribution to the Properties of Carbon Fibre Reinforced Composites. H.W. Bergmenn. ESA-TT-849, July 1984.

Cracks in Composite Materials. Vol. 6 of *Mechanics of Fracture*. G.C. Sih and E.P. Chen. The Hague: Martinus Nijhoff Publishers, 1981.

"Critical Look at Current Applications of Fracture Mechanics to the Failure of Fibre-Reinforced Composites." M.F. Kanninen, E.F. Rybicki, and H.F. Brinson. *Composites* 18:17–22, 1977.

"Damage Accumulation and Residual Strength Degradation in Graphite/Epoxy Laminates: Failure Modes and NDE of Composites." I.M. Daniel and A. Charewicz. Paper no. 12 presented at TTCP Workshop, July 1984.

"Damage and Failure Analysis of Composite Structures with Manufacturing Defects." R. Talreja. Paper presented at ECCM15, 15th European Conference on Composite Materials, Venice, June 2012.

"Damage Behavior of Composite Structures and Joints at Room and Elevated Temperatures." F. Delale, A.D. Walser, and B.M. Liaw. DAAE07-96-C-X-121, 1998.

Damage in Composite Materials: Basic Mechanisms, Accumulation, Tolerance, and Characterization, edited by K.L. Reifsnider. ASTM STP 775, 1982.

Damage Mechanics of Composite Materials: Constitutive Modeling and Computational Algorithms. J.W. Ju. AFOSR-89-0020, 1991.

"Damage Mechanisms in CFRP Laminates with Defects: Contributions on the Properties of Carbon Fibre Reinforced Composites." L. Kirschke. In ESA-TT-849, 155–204, July 1984.

Damage Tolerance in Advanced Composite Materials. G. Dorey. RAE-TR-77172, November 1977.

Damage Tolerance of Carbon Fibre Reinforced Composites, The: A Workshop Summary. K.D. Challenger. ONRL Rpt. C-15-85, January 1986.

Damage Tolerance of Composites: Changing Emphasis in Design, Analysis and Testing. D.J. Wilkins. General Dynamics, Fort Worth, TX, January 1983.

"Damage Tolerance of Graphite/Epoxy Composites." A.A. Baker, R. Jones, and R.J. Callinan. *Composite Structures* 4:15–44, 1985.

Damage Tolerance of Sandwich Structures. R.C. Moody and A.J. Vizzini. DOT-FAA-AR-99-91, U.S. Department of Transportation Federal Aviation Administration Office of Aviation Research, 2000.

"Damage/Defect Types and Inspection: Some Regulatory Concerns." S. Waite. Paper presented at MIL-17 Damage Tolerance and Maintenance Workshop, Chicago, July 2006.

Defect Criticality in Composite Structures: Delaminations.
K.N. Street. Interim report submitted to TTCP Subgroup
P: Materials TTP 4, April 1987.

"Defect Occurrences in the Manufacture of Large CFC Structures
and Work Associated with Defects, Damage and Repair of
CFC Components." C.S. Frame and G. Jackson. Chap. 21 in
*Characterization, Analysis and Significance of Defects in
Composite Materials*, AGARD-CP-335, April 1983.

"Defect Pattern Recognition by Sensor Data Fusion in the 'Coin
Tap Test'." H. Wu and M. Siegel. Powerpoint presentation.
The Robotics Institute, SCS Carnegie Mellon University.

"Defect/Damage Tolerance of Pressurized Fiber Composite Shells."
C.C. Chamis and L. Minnetyan. *Composite Structures* 51:159–68,
2001.

"Defects in Composite Materia Caused by Drilling in Manufacturing
Process." M. Colt-Stoica, D. Anania, C. Mohora, and D. Stoica.
Manufacturing and Industrial Engineering 2011 (3): 27–29,
2011. http://web.tuke.sk/fvtpo/journal/pdf11/3-pp-27-29.pdf.

*Defects of Importance in the Specification of Reinforced Plastics
Products.* F.R. Barnst. NAVORD Report 279, 1953.

*Deformation, Constitutive Behavioral Damage of Advanced
Structural Materials under Multi-Axial Loading.* S.D.
Antolovich. AFOSR-90-0162, 1993.

*Degradation of Graphite/Epoxy Composite Materials Because of
Load Induced Micromechanical Damage.* G.C. Grimes and
J.M. Whitney. Southwest Research Inst. San Antonio, TX, 1974.

"Delamination and Failure at Ply Drops in Carbon Fiber Laminates
under Static and Fatigue Loading." D.D. Samborsky, D.P.
Avery, P. Agastra, and J.F. Mandell. Paper presented at 44th
AIAA Aerospace Sciences Meeting and Exhibit, Reno, Nevada,
January 2006.

"Delamination Crack Growth in Composite Laminates: Delamination
and Debonding of Materials." A.S.D. Wang, M. Slominana, and
R.B. Bucinell. In *Delamination and Debonding of Materials*,
edited by W.S. Johnson, 135–67. ASTM STP 876, 1985.

"Describe NDI Techniques Currently Available in the Field."
In *Describe Composite Damage and Repair Inspection
Procedures*, Section H1, TCO H Module. http://www.niar.
wichita.edu/niarworkshops/Portals/0/Mod%20H_rev_4.1.pdf.

*Design Methodology for Single Sided Supported Repairs to Through
Thickness Defects.* AAP 7021.016-1 Ch06c2-0.01D, October 2005.

"Design of Repairable Advanced Composite Structures, The."
L.J. Hart-Smith. Douglas Paper 7550, October 1985.

"Determining Fracture Directions and Fracture Origins on Failed
Graphite/Epoxy Surfaces: Non-Destructive Evaluation
and Flaw Criticality for Composite Materials." G.E. Morris.
In *Nondestructive Evaluation and Flaw Criticality for
Composite Materials*, edited by R.B. Pipes, W.R. Scott,
S.V. Kulkarni, and W.W. Stinchcomb, 274–97. ASTM STP
696, 1979.

Developments in Reinforced Plastics, edited by G. Pritchard. Vol.
2, *Properties of Laminates*, Chapter 2, "Imperfections in FRP
Materials." London: Applied Science, 1982.

"Distribution of Defects in Wind Turbine Blades and Reliability
Assessment of Blades Containing Defects." H. Stensgaard Toft,
K. Branner, P. Berring, and J.D. Sørensen. EWEC 2009 Scientific
Proceedings, EWEC, 42–47, 2009.

"D-Sight Technique for Rapid Impact Damage Detection on
Composite Aircraft Structures." J.H. Heida and A.J.A. Bruinsma.
Paper presented at ECNDT '98, 1998.

"Effect of Defects on Aircraft Composite Structures."
R.A. Garrett. Chap. 19 in *Characterization, Analysis
and Significance of Defects in Composite Materials*.
AGARD-CP-335, April 1983.

"Effect of Flaws and Porosity on Strength of Adhesive Bonded
Joints." L.J. Hart-Smith. Douglas Paper 7388, presented to 29th
Annual SAMPE Symposium and Tech. Conf., April 1984.

"Effect of Geometry on Interlaminar Stresses of [0/90]ₛ Composite
Laminates with Circular Holes, The." W.M. Lucking, S.V. Hoa,
and T.S. Sankar. Presented at DND Composite Materials
Workshop, DREP-SP-83-1, 15.1, July 1983.

"Effect of Geometry on the Mode of Failure of Composites in
Short-Beam Shear Test, The." In *Composite Materials: Testing
and Design*, edited by S.A. Sattar and D.H. Kellogg, 62–71.
ASTM STP 460, 1969.

*Effect of Impact Damage and Open Holes on the Compression
Strength of Tough Resin/High Strain Fibre Laminates.*
J.G. Williams. NASA-TM-85756. February 1984.

"Effect of Low-Velocity Impact Damage on the Fatigue
Behaviour of Graphite/Epoxy Laminates." R.L. Ramkumar.
In *Long-Term Behaviour of Composites*, 116–35. ASTM STP
813, 1983.

"Effect of Manufacturing Defects and Service-Induced Damage on the Strength of Aircraft Composite Structures." R.A. Garrett. In *Composite Materials: Testing and Design* (7th Conf.), 5–33. ASTM STP 893, 1986; and "Composite Repairs." In *SAMPE Monograph No. 1*, 208–24. Covina CA: SAMPE.

"Effect of Matrix Toughness on Delamination, The: Static and Fatigue Fracture under Mode I Shear Loading of Graphite Fibre Composites." A.J. Russell and K.N. Street. Paper presented at NASA/ASTM Symposium on Composites, Houston, TX, March 1985.

"Effect of Stacking Sequence on Damage Accumulation." D.B. Runckle and A.B. Doucet. Paper presented at 7th ASC Conference, 1992.

"Effects and Structural Significance of Defects and Damage in Composite Materials, The." R.T. Potter and G. Dorey. RAE, Farnborough, 1984.

"Effects of Bolt-Hole Contact on Bearing-Bypass Damage-Onset Strength." J.H. Crews Jr. and R.A. Naik. AIAA Paper, 1991.

"Effects of Defects: Part A; Development of a Protocol for Defect Risk Management and Improved Reliability of Composite Structures." T.W. Riddle, D.S. Cairns, J.W. Nelson, and J. Workman. Paper presented at AIAA Structures, Dynamics, and Materials Conference, 2012. http://www.coe.montana.edu/composites/documents/2012_AIAA_2012 1420%20Riddle.pdf.

"Effects of Defects and Damage in Composites." M. Ratwani. Private communication, 2005.

Effects of Defects in Composite Materials: A Symposium. D. Wilkins. ASTM Special Technical Publication 836 STP, 1982.

"Effects of Material and Stacking Sequence on Behavior of Composite Plates with Holes." I.M. Daniel, R.E. Rowlands, and J.B. Whiteside. *Experimental Mechanics* 14 (1): 1–9, 1974.

Effects of Materials and Processes Defects on the Final Report Compression Properties of Advanced Composites. R.L. Ramkumar, G.C. Grimes, D.F. Adams, and E.G. Dusablon. NOR 82-103, September 1980 to February 1982.

"Effects of Porosity, Delamination and Low Velocity Impact Damage on the Compressive Behaviour of Graphite/Epoxy Laminates." R.L. Ramkumar. In *Symposium on NDE of Criticality of Defects in Composite Laminates*, 305–56. NADC-84041-60, May 1983.

Engineering Materials Handbook. Vol. 1, *Composites*. Metals Park, OH: ASM Int., 1988.

"Engineering Significance of Defects in Composite Structures, The."
D.J. Wilkins. In *Characterization, Analysis and Significance of
Defects in Composite Materials*. AGARD-CP-335, April 1983.

Environmental Durability of Graphite/Epoxy Composites. R. Chester.
Commonwealth Aeronautical Advisory Research Committee
(CAARC) Composites Workshop. Aeronautical Research
Laboratories, Melbourne, May 1988.

*Environmental Effects on the Post-Impact Compressive Strength of
CFC*. R.T. Potter. RAE TM MAT/STR 1088, 1987.

Evaluation of a Damaged F/A-18 Horizontal Stabilator. J. Paul. ARL
Struc. TM 590, February 1989.

"Evaluation of Damage Analysis Techniques for Composite Aircraft
Structures—Executive Summary." R.B. Heslehurst. Defence
Fellowship Report, Australian Defence Department, February
1990.

"Evaluation of Flawed Composite Structure under Static and Cyclic
Loading." T.R. Potter. In *Fatigue of Filamentary Composite
Materials*, 152–70. ASTM STP 636, 1977.

"Evaluation of Stress Intensity Factors of Composite Laminates
Using Finite Element Techniques." B. Hansen. Presented at
DND Composite Materials Workshop, DREP-SP-83-1, 16.1, July
1983.

"Expanding the Capabilities of the Ten-Percent Rule for Predicting
the Strength of Fibre-Polymer Composites." L.J. Hart-Smith.
In *Failure Criteria in Fibre-Reinforced-Polymer Composites*,
edited by M.J. Hinton, A.S. Kaddour, and P.D. Soden, 597–642.
London: Elsevier Ltd., 2004.

*Experimental Aspects of Using Time-Averaged Holographic
Interferometry to Detect Barely Visible Impact Damage
in a Graphite/Epoxy Composite Plate*. S.J. Rumble. ARL-
STRUC-TM-467, 1987.

"Factors Affecting the Interlaminar Fracture Energy of Graphite/
Epoxy Laminates." A.J. Russell and K.N. Street. Paper pre-
sented at Progress in Science and Engineering of Composites,
ICCM-IV, Tokyo, 279–86, 1982.

*Factors Affecting the Opening Mode Delamination of Graphite/Epoxy
Laminates*. A.J. Russell. DREP Materials Rpt. 82-Q, December
1982.

"Failure Analysis and Mechanisms of Failure of Fibrous Composite
Structures." Compiled by A.K. Noor, S.G. Williams, M.J. Shuart,
and J.H. Starnes Jr. In *Proc. of a Workshop on Failure Analysis*

and Mechanisms of Failure of Fibrous Composite Structures. NASA-CP-2278, March 1982.

Failure Analysis of a Graphite/Epoxy Laminate Subjected to Bolt Bearing Loads. J.H. Crews Jr. and R.V.A. Naik. NASA-TM-86297, August 1984.

"Failure Analysis of Composites with Stress Gradients." E.M. Wu. In *Proc. of 1st USA-USSR Symposium on Fracture of Composite Materials*, 63–76, September 1978.

Failure Characteristics of Graphite-Epoxy Structural Components Loaded in Compression. J.H. Starnes Jr. and J.G. Williams. NASA-TM-84552, September 1982.

Failure Initiation and Effect of Defects in Structural Discontinuous Fiber Composites. B. Boursier and A. Lopez. Society for the Advancement of Material and Process Engineering (SAMPE), 2010. http://hexcel.com/Innovation/Documents/Failure%20 Initiation%20and%20Effect%20of%20Defects%20in%20 Structural%20Discontinuous%20Fiber%20Composites.pdf.

"Failure Mechanisms and Fracture of Composite Laminates with Stress Concentrations." I.M. Daniel. In *Proc. 7th Int. Conf. on Experimental Stress Analysis*, 1–20, August 1982.

"Failure Mode and Strength Predictions of Mechanically Fastened Composite Joints." Z. Maekawa, A. Kaji, H. Hamada, and M. Nagamori. In *Proc. 5th Int. Conf. on Composite Materials*, 99–109, July 1985.

Failure Modes of Fibre Reinforced Laminates. ESDU 82025 Amend. A, June 1986.

"Failure of Materials in Mechanical Design: Analysis, Prediction, Prevention." J.A. Collins. Chap. 2 in *Modes of Mechanical Failure.* New York: John Wiley and Sons, 1981.

"Failure of Quasi-Isotropic Composite Laminates with Free Edges." C.T. Sun and S.G. Zhou. *J. Reinforced Plastics and Composites* 7:515–57, 1988.

Failure of Sandwich Structures with Sub-Interface Damage. A. Shipsha. Royal Institute of Technology, Stockholm, Report 2001-13, May 2001. https://pubweb.bnl.gov/~e865/KOPIO/ Vacuum/Sandwich_structure.pdf.

"Fast Interlaminar Fracture of a Compressively Loaded Composite Containing a Defect." M. Ashizawa. Douglas Paper 6994, presented to Fifth DoD/NASA Conference on Fibrous Composites in Structural Design, New Orleans, Louisiana, January 1981.

Fatigue Damage–Strength Relationships in Composite Laminates.
K.L. Reifsnider, W.W. Stinchcomb, E.G. Henneke, and J.C.
Duke. AFWAL-TR-S3-3084-V2, January 1984.

"Finite Element Analysis of Damage in Fibrous Composites Using
a Micromechanical Model." J.M. Bemer. Naval Postgraduate
School, December 1993.

"Flaw Criticality of Graphite/Epoxy Structures: Nondestructive
Evaluation and Flaw Criticality for Composite Materials." D.Y.
Konishi and K.II. Lo. In *Nondestructive Evaluation and Flaw
Criticality for Composite Materials*, edited by R.B. Pipes, W.R.
Scott, S.V. Kulkarni, and W.W. Stinchcomb, 125–44. ASTM STP
696, 1979.

"Fractographical Investigations." P. Nitsch. In *Mechanical
Properties and Damage Mechanisms of Carbon
Fiber-Reinforced Composites: Tension Loading*. DFVLR-FB
85-45, June 1985.

"Fractographic Analysis of Failures in CFRP." D. Purslow. In
*Characterization, Analysis and Significance of Defects in
Composite Materials*, 1-1. AGARD-CP-355, April 1983.

"Fracture Mechanics for Delamination Problems in Composite
Materials." S.S. Wang. In *Progress in Science and Engineering
of Composites*, 287–96. Proc. 4th Int. Conf. on Composite
Materials, Tokyo, October 1982.

"Fracture Mechanics of Composite Materials." G.C. Sih. In *Proc.
1st USA-USSR Symposium on Fracture of Composite Materials*,
111–30, September 1978.

"Fracture Mechanics Methodology: Evaluation of Structural
Components Integrity." G.C. Sih. Chap. 2 in *Fracture Mechanics
of Engineering Structural Components*. The Hague: Martinus
Nijhoff, 1984.

"Fracture of Thick Laminated Composites." C.E. Harris and D.H.
Morris. *Experimental Mechanics* 26:34–41, 1986.

*Fracture Surface Characteristics of Notched Angled Plied Graphite/
Epoxy Composites.* C.A. Ginty and T.B. Irvine. NASA-TM-83786,
September 1984.

"Fracture Toughness of Fibre Composites, The." D.C. Phillips and
A.S. Tetelman. *Composites* 3 (5): 216–23, 1972.

"Growth Mechanisms of Transverse Cracks and Ply Delamination in
Composite Laminates." A.S.D. Wang. In *Advances in Composite
Materials*, 170–85. Proc. 3rd Int. Conf. on Composite Materials,
Paris, August 1980.

Guidelines for Analysis, Testing, and Nondestructive Inspection of Impact Damaged Composite Sandwich Structures. P. Shyprykevich, J. Tomblin, L. Ilcewicz, A.J. Vizzini, T.E., Lacy, and Y. Hwang. DOT-FAA-AR-02-121, March 2003. http://www.tc.faa.gov/its/worldpac/techrpt/ar02-121.pdf.

"High Performance Fibre Composites for Aircraft Applications: An Overview." A.A. Baker. *Metals Forum* 6 (2): 81–101, 1983.

Holographic Interferometry Assessment in Composite and Adhesively Bonded Structural Defects. R. Heslehurst. Saarbrücken, Germany: Lambert Academic Publishing, 2010.

"Holographic Interferometry Study of Free Edge Defects." Nathan Gilmore. Undergraduate Thesis, University College, UNSW, 2006.

"Holographic Techniques for Defect Detection." G.E. Maddux and G.P. Sendeckyj. In *Nondestructive Evaluation and Flaw Criticality for Composite Materials*, edited by R.B. Pipes, W.R. Scott, S.V. Kulkarni, and W.W. Stinchcomb, 26–44. ASTM STP 696, 1979.

Hygrothermal Damage Mechanisms in Graphite-Epoxy Composites. F.W. Crossman, R.E. Mauri, and W.J. Warren. NASA-CR-3189, December 1979.

"Hygrothermal Effects on the Initiation and Propagation of Damage in Composite Shells." A. Ghosh. *Aircraft Engineering and Aerospace Technology* 80 (4): 386–99, 2008.

"Impact Damage and Energy-Absorbing Characteristics and Residual In-Plane Compressive Strength of Honeycomb Sandwich Panels." G. Zhou and M.D. Hill. *Journal of Sandwich Structures and Materials* 11 (4): 329–56, 2009.

"Impact Damage in Composites: Development, Consequences and Prevention." G. Dorey. In *Proc. 6th Int. Conf. on Composite Materials*, 3.1–3.26, 1987.

"Improving In-Service Inspection of Composite Structures: It's a Game of Catt and Maus." D. Roach and K. Rackow. Paper presented at DoD/NASA/FAA Aging Aircraft Conference, 2003.

"Indentation Failure in Composite Sandwich Structures." E.E. Gdoutos, I.M. Daniel, and K.A. Wang. *Experimental Mechanics* 42 (4): 426–31, 2002.

"Influence of Damage on the Elastic Behaviour of Composite Beams." A. Makeev and E.A. Armanios. American Helicopter Society Annual Form, Conference Paper, 1999.

"Influence of Prescribed Delaminations on Stiffness-Controlled Behaviour of Composite Laminates." A.D. Reddy, L.W. Rehfield, and R.S. Haag. In *Effects of Defects in Composite Materials*, 71–83. ASTM STP 836, 1984.

"Initiation and Growth of Transverse Cracks and Edge Delamination in Composite Laminates, Part 1: An Energy Method." A.S.D. Wang and F.W. Crossman. *J. Composite Materials Suppl.* 14:71–87, 1980.

"In-Plane Tensile Strength of Multidirectional Composite Laminates." R.Y. Kim. In *Progress in Science and Engineering of Composites*, 455–64. Proc. 4th Int. Conf. on Composite Materials, Tokyo, October 1982.

"In-Service NDI of Composite Structures: An Assessment of Current Requirements and Capabilities." D.E.W. Stone and B. Clarke. Chap. 18 in *Characterization, Analysis and Significance of Defects in Composite Materials*. AGARD-CP-335, April 1983.

Integrity Control of CFRP Structural Elements: Phase 1 Rpt. W.H. Paul and D. Wagner. ESA-CR(P)-1778, April 1983.

"Interaction of Impact Damage and Tapered-Thickness Sections in CFRP, The." R.T. Potter. *Composite Structures* 3:319–39, 1985.

Interim Report on Environment Program: Durability of Graphite/ Epoxy Honeycomb Specimens with Representative Damage and Repairs. P.D. Chalkley and R.J. Chester. ARL Aircraft Materials TM 397, July 1988.

"Interlaminar Defect Criticality for Compressive Loading." R.B. Pipes, J.D. Webster, and W.A. Dick. *Progress in Science and Engineering of Composites, ICCM-IV* 1 (4): 269–70, 1982.

"Interlaminar Fracture of Composite Materials." R.A. Jurf and R.B. Pipes. *J. Composite Materials* 16:386–94, 1982.

"Investigation of Fastener Hole Delaminations." M. Betts. Undergraduate Thesis, University of New South Wales (UNSW), Australia, October 1995.

"Is Fatigue Testing of Impact Damaged Laminates Necessary?" R. Jones, J.F. Williams, and T.E. Tay. *Composite Structures* 8:1–12, 1987.

Lectures on Composite Materials for Aircraft Structures. B.C. Hoskin and A.A. Baker, ed. ARL Struc. Rpt. 394, October 1982.

"Location and Direction of the Propagation of an Existing Delamination in Composite Panels." R.B. Heslehurst, J.P. Baird, and H.M. Williamson. *Fatigue and Fracture of Engineering Materials and Structures* 1996.

"Low Velocity Impact Damage in Carbon Fiber Reinforced Plastic Laminates." W.J. Cantwell and J. Morton. In *Proc. of the 5th Int. Congress on Experimental Mechanics*, 314–19. Montreal, June 1984.

"Manufacturing Defects Common to Composite Wind Turbine Blades: Effects of Defects." J.W. Nelson, D.S. Cairns, and T.W. Riddle. Paper presented at 52nd AIAA/ASME/ASCE/AHS/ASC Structures, Structural Dynamics and Materials Conference, April 2011, Denver, CO.

"Manufacturing Defects in Fiber Reinforced Plastics Composites." J. Summerscales. *Insight* 36 (12): 936–942, 1994.

"Matrix Deformation and Fracture in Graphite Reinforced Epoxies." W.L. Bradley and R.N. Cohen. In *Delamination and Debonding of Materials*, 389–410. ASTM STP 876, 1985.

Mechanical Behaviour and Fracture Characteristics of Off-Axis Fiber Composites, I: Experimental Investigation. J.H. Sinclair and C.C. Chamis. NASA-TP-1081, December 1977.

Mechanical Properties and Damage Mechanisms of Carbon Fiber-Reinforced Composites: Tension Loading. H.W. Bergmenn. DFVLR-FB 85-45, 343, June 1985.

"Mechanistic Failure Criteria for Carbon and Glass Fibers Embedded in Polymer Matrices." L.J. Hart-Smith. Boeing Paper MDC 01K0102 AIAA Paper AIAA-2001-1183, presented to 42nd AIAA/ASME/ASCE/AHS/ASC Structures, Structural Dynamics, and Materials Conference, Seattle, WA, April 2001.

"Mechanistic Failure Criteria for Polymer Matrices Constrained between Carbon and Glass Fibers." J.H. Gosse and S. Christensen. AIAA Paper AIAA-2001-1184, presented to 42nd AIAA/ASME/ASCE/AHS/ASC Structures, Structural Dynamics, and Materials Conference, Seattle, WA, April 2000.

"Methodology for Composite Durability Assessment." S.W. Tsai and J.L. Townsley. Paper presented at SAMPE Technical Conference, Dayton, OH, 2003.

"Mixed-Mode Fracture Analysis of $(+/-25/90_n)_s$ Graphite/Epoxy Composite Laminates (A)." G.E. Law. In *Effect of Defects in Composite Materials*, 143–60. ASTM STP 836, 1984.

"MOD Funded Programs on Damage Tolerance of Composite Structures." R.T. Potter. Commonwealth Aeronautical Advisory Research Committee (CAARC) Composites Workshop. Aeronautical Research Laboratories, Melbourne, May 1988.

Modeling Considerations in the Prediction of Residual Strength in Composite Laminates. E. Saether. ARL-MR-211, PDF, November 1994.

NDI Survey of Composite Structures. S.A. McGovern. NADC-80032-60, May 1979.

"New Ply Model for Interlaminar Stress Analysis (A)." R.R. Valisetty and L.W. Rehfield. In *Delamination and Debonding of Materials*, 52–68. ASTM STP 876, 1985.

Nondestructive Evaluation and Flaw Criticality for Composite Materials, edited by R.B. Pipes, W.R. Scott, S.V. Kulkarni, and W.W. Stinchcomb. ASTM STP 696, 1979.

Nondestructive Flaw Definition Techniques for Critical Defect Determination: Final Report. F. Sattler. NASA-CR-72602, September 1970.

Nondestructive Testing of Fibre-Reinforced Plastics Composites, Vol. 1, edited by J. Summerscales. London: Elsevier Applied Science, 1987.

Nondestructive Testing of Honeycomb Assemblies and Composite Structures. McDonnell Aircraft Co., Process Specification PS-21233, Rev. D Suppl. 1A, October 1972.

Nondestructive Testing Personnel Qualification and Certification (Eddy Current, Liquid Penetrant, Magnetic Particle, Radiographic and Ultrasonic). MIL STD 410 Revision E. 31, December 1997.

"Notch Sensitivity and Stacking Sequence of Laminated Composites." P.A. Lagace. In *Composite Materials: Testing and Design* (7th Conf.), edited by J.M. Whitney, 161–76. ASTM STP 893, 1986.

"Numerical Modeling and Experimental Analysis of Subsurface Defects in Composite Panels." A. Muc and P. Pastuszak. Mechanics of Nano, Micro and Macro Composite Structures Politecnico di Torino, June 2012.

"Numerical Simulation of Impact Damage in Composite Laminates." R. Banerjee. Paper presented at 7th ASC Conference, 1992.

"Observations in the Structural Response of Adhesive Bondline Defects." R.B. Heslehurst. *International Journal of Adhesion & Adhesives* 19:133–54, 1999.

"On Failure Modes in Finite Width Angle Ply Laminates." C.T. Herakovich. In *Advances in Composite Materials*, 425–35. Proc. 3rd Int. Conf. on Composite Materials, Paris, August 1980.

"On Failure Modes of Unidirectional Composites under Compressive Loading." L.B. Greszczuk. In *Fracture of Composite Materials,* 231–44. Proc. 2nd USA-USSR Symposium, Bethlehem, PA, March 1981.

"On Mixed Mode Fracture in Off-Axis Unidirectional Graphite/ Epoxy Composites." A.S.D. Wang, N.N. Kishore, and W.W. Feng. In *Progress in Science and Engineering of Composites,* 599–606. Proc. of the 4th Int. Conf. on Composite Materials, Tokyo, October 1982.

"On the Monitoring of Degradation of Composite Materials Using Pattern Recognition Method." S.S. Tang, K.L. Chen, and J.E. Grady. Paper presented at 7th ASC Conference, 1992.

"On the Three Dimensionality of Failure Modes in Angle-Ply Strips under Tension." D.W. Oplinger, B.S. Parker, and A. Grenis. In *Composite Materials: The Influence of Mechanics of Failure on Design,* 263–86. Proc. Army Symposium on Solid Mechanics, AMMRC MS 76-2, September 1976.

"On the Underrated Value of Visual Inspections for Adhesively Bonded and Fibrous Composite Structures." J.L. Hart-Smith. *Science of Advanced Materials & Process Engineering* 40 (2): 1134–45, 1995.

"Overview of Composite Damage and Structural Repair." R.J. Ducar. Paper presented at SAMPE Conference, May 2000.

Polymer Matrix Composite (PMC) Damage Tolerance and Repair Technology. R.Y. Kim. AFRL-ML-WP-TR-2001-4176, April 2001.

"Prediction of Threshold Loads for Long Fatigue Life of Sandwich Structures with Wrinkle Defects." M. Leong, L.C.T. Overgaard, O.T. Thomsen, and E. Lund. Paper presented at 16th International Conference on Composite Structures (ICCS 16), 2011.

"Predictions of a Generalized Maximum-Shear-Stress Failure Criterion for Certain Fibrous Composite Laminates." L.J. Hart-Smith. In *Failure Criteria in Fibre-Reinforced-Polymer Composites,* edited by M.J. Hinton, A.S. Kaddour, and P.D. Soden, 219–263. London: Elsevier Ltd., 2004.

"Predictions of the Original and Truncated Maximum Strain Failure Models for Certain Fibrous Composite Laminates." L.J. Hart-Smith. In *Failure Criteria in Fibre Reinforced Polymer Composites,* edited by M.J. Hinton, A.S. Kaddour, and P.D. Soden, 179–218. London: Elsevier Ltd., 2004.

Predictive Modelling of Damage in Composite Materials. W.T. Chester, S.E. Fielding, and P.D. Hilton. British Ministry of Defence under Contract No. A91A/1378, March 1985.

"Preliminary Damage Tolerance Methodology for Composite Structures (A)." D.J. Wilkins. In *Proc. of a Workshop on Failure Analysis and Mechanisms of Failure of Fibrous Composite Structures*, 67–94. NASA-CP-2278, March 1982.

"Preliminary Development of a Fundamental Analysis Model for Crack Growth in a Fiber Reinforced Composite Material." M.F. Kanninen, E.F. Rybicki, and W.I. Griffith. In *Composite Materials: Testing and Design* (4th Conf.), 53–69. ASTM STP 617, 1977.

"Principles for Achieving Damage Tolerant Primary Composite Aircraft Structures." H. Razi and S. Ward. Paper presented at 11th DoD/FAA/NASA Conf. on Fibrous Composites in Structural Design, Fort Worth, TX, August 1996.

Probabilistic Design of Damage Tolerant Composite Aircraft Structures. A. Ushakov, A. Stewart, I. Mishulin, and A. Pankov. DOT-FAA-AR-01-55, 2002.

"Progressive Damage and Strength of Bolted Joints in Composite Structures." F.-X. Irisarri, F. Laurin, N. Carrère, and J.F. Maire. *Composite Structures* 2012. http://extra.ivf.se/eccm13_programme/abstracts/1108.pdf.

Progressive Fracture in Composites Subjected to Hygrothermal Environment. L. Minnetyan, P.L.N. Murthy, and C.C. Chamis. NASA-TM-105230, April 1991.

"Progressive Transverse Cracking and Local Delamination in Composite Laminates." J.R.D. Runkle. In *Proceedings of the American Society for Composites (ASC) Conference*, 276–85, 1992.

"Propagation of Hole Machining Defects in Pin-Loaded Composite Laminates." E. Persson, I. Eriksson, and P. Hammersberg. *Journal of Composite Materials* 31 (4): 383–408,1997.

Recognizing Defects in Carbon-Fibre Reinforced Plastics. R. Schnetze and W. Hillger. NASA-TM-76947, September 1982.

Repair of Composites: Composite Structure Repair. T.M. Donnellan, E.L. Rosenzweig, R.E. Trabocco, and J.G. Williams. AGARD Report No. 716 Addendum, August 1984.

Repair of Damage to Marine Sandwich Structures Part I: Static Testing. R. Thomson, R. Luescher, and I. Grabovac. DSTO-TR-0736, May 2000.

Repair of Damage to Marine Sandwich Structures Part II: Fatigue Testing. M.Z. Shah Khan and I. Grabovac, DSTO-TN-0275, May 2000.

Residual Strength of Composites with Multiple Impact Damage. J.J. Paul, S.C. Galea, and R. Jones. ARL-RR-13, March 1994.

"Residual Strength of Impact Damaged Composites." R. Jones, A.A. Baker, and R.J. Callinan. *Int. J. Fracture* 24:R51, 1984.

"Review of Defects and Damage Pertaining to Aircraft Composite Structures." R.B. Heslehurst and M.L. Scott. *J. Composite Polymers* 3 (2): 103–33, 1990.

"Review of Low-Velocity Impact Properties of Composite Materials." M.G.W. Richardson and M.J. Wisheart. *Composites Part A: Applied Science and Manufacturing* 27 (12): 1123–31, 1996.

"Selective Generation of Lamb Wave Modes and Their Propagation Characteristics in Defective Composite Laminates." Z. Su and L. Ye. *Proceedings of the Institution of Mechanical Engineers, Part L: Journal of Materials Design and Applications* 218 (2): 95–110, 2004.

"Shear Properties of a Sandwich Composite Containing Defects." R.S. Thomson, M.Z. Shah Khanb, and A.P. Mouritzb, *Composite Structures* 42:107–18, 1998.

Significance of Defects in Composite Structures, The: A Review of Current Research and Requirements. G. Dorey, R.F. Mousley, and R.T. Potter. RAE TM Mat/Str-1014, April 1983.

Significance of Defects on the Failure of Fibre Composites, The. S.M. Bishop. RAE-TM-MAT-394, March 1982.

"Simple Model to Simulate the Interlaminar Stresses Generated near the Free Edge of a Composite Laminate (A)." P. Conti and A. De Paulis. In *Delamination and Debonding of Materials*, 35–51. ASTM STP 876, November 1983.

"Simulation of Mechanical Behaviour of Composite Bonded Joints Containing Strip Defects." M.F.S.F. de Mouraa R. Daniauda, and A.G. Magalhaesb. *International Journal of Adhesion & Adhesives* 26:464–73, 2006.

"Some Physical Defects Arising in Composite Material Fabrication: Review." W. Johnson and S.K. Ghosh. *J. Materials Science* 16:285–301, 1981.

States of Stress and Strain in Adhesive Joints, Including Photoelastic Imaging of Defects in Adherends. J.P. Sargent, T.W. Turner, and K.H.G. Ashbee. AFOSR-TR-79-1359, Report, November 1979.

"Stiffness-Reduction Mechanisms in Composite Laminates."
A.L. Highsmith and K.L. Reifsnider. In *Damage in Composite Materials: Basic Mechanisms, Accumulation, Tolerance, and Characterization*, 103–117. ASTM STP 775, 1982.

"Strain and Stress Concentrations in Composite Laminates Containing a Hole." S.C. Tan and R.Y. Kim. *Experimental Mechanics* 30 (4): 345–351, 1990.

"Strength of Quasi-Isotropic Laminates under Off-Axis Loading." S.R. Swanson and B.C. Trask. *Composites Science and Technology* 34:19–34, 1989.

Structural Degradation in Fibre Composites by Kinking. A.G. Evans and W.F. Adler. Rockwell Int. Science Centre, Thousand Oaks, CA, May 1977.

Structural Significance of Failure Mode in Notched Fibre Reinforced Plastics under Tension, The. R.T. Potter. RAE TR 82009, January 1982.

"Studies on Mechanical Behavior of Glass Epoxy Composites with Induced Defects and Correlations with NDT Characterization Parameters." D. Pradeep, N. Janardhana Reddy, C.R. Kumar, L. Srikanth, and R.M.V.G.K. Rao. *Journal of Reinforced Plastics and Composites* 26 (15): 1539–56, 2007.

"Technological Defects Classification System for Sandwiched Honeycomb Composite Materials Structures." A.V. Gaydachuk, M.B. Slivinskiy, and V.A. Golovanevskiy. *Materials Forum* 30:96–102, 2006. http://www.materialsaustralia.com.au/lib/pdf/Mats.%20Forum%20page%2096_102.pdf.

"The Effect of Defects on the Performance of Post-Buckled CFRP Stringer-Stiffened Panels." E. Greenhalgh, C. Meeks, A. Clarke, and J. Thatcher. *Composites Part A: Applied Science and Manufacturing* 34 (7): 623–33, 2003.

"The Effects of Fastener Hole Defects." S.D. Andrews, O.O. Ochoa, and S.D. Owens. *Journal of Composite Materials* 27 (1): 2–20, 1993.

The Impact of Accommodating Defects on the Efficiency of Aircraft Design. E.M. Petrushka. General Dynamics, Fort Worth Division, TX.

The Inclusion of In-Plane Stresses in Delamination Criteria. M.T. Fenske and A.J. Vizzini. AIAA-98-1743, 1998.

"The Influence of Defects and Damage on the Strength of FRP Sandwich Panels for Naval Ships." B. Hayman and D. Zenkert. Paper presented at PRADS 2004, Travemünde, Germany, 2004.

The Significance of Defects in CFRP Bonded Honeycomb Structures and Non-Destructive Test Methods for Their Destruction. R.L. Crocker and W.H. Bowyer. Fulmer Research Labs Stoke Poges, U.K. For European Space Agency R736/4, May 1978.

"Theoretical and Experimental Analysis of a Tensile Member of Anisotropic Material under Static and Fluctuating Loads (A)." G. Hedley and B.S. Owen. Paper 5 in *Fibre Reinforced Materials*. London: Institute of Civil Engineers, 1977.

"Three-Dimensional Study of Delamination (A)." E. Altus and A. Dorogoy. *Engineering Fracture Mechanics* 33 (1): 1–19, 1989.

"Ultrasonic NDE of Fibre Reinforced Composite Materials: A Review." V.K. Kinra and V. Dayal. Academy Proc. in Engineering Science, Indian Academy of Science 11 (3–4): 419–432, December 1987.

"Uniaxial Failure of Composite Laminates Containing Stress Concentrations." R.J. Nuismer and J.M. Whitney. In *Fracture Mechanics of Composites*, 117–42. ASTM STP 593, 1975.

"Visual Inspection of Carbon Epoxy Composite Parts." McDonnell Aircraft Co. Process Specification P.S. 21252, February 1984.

Index

Printed and bound by CPI Group (UK) Ltd, Croydon, CR0 4YY

18/10/2024

01776271-0004